CONCEPTS IN PLANT PHYSIOLOGY

CONCEPTS IN PLANT PHYSIOLOGY

Dr. Sarfaraz Ahmad

ANMOL PUBLICATIONS PVT. LTD.
NEW DELHI-110 002 (INDIA)

ANMOL PUBLICATIONS PVT. LTD.
Regd. Office: 4360/4, Ansari Road, Daryaganj,
New Delhi-110002 (India)
Tel.: 23278000, 23261597, 23286875, 23255577
Fax: 91-11-23280289
Email: anmolpub@gmail.com
Visit us at: www.anmolpublications.com

Branch Office: No. 1015, Ist Main Road, BSK IIIrd Stage
IIIrd Phase, IIIrd Block, Bengaluru-560 085 (India)
Tel.: 080-41723429 • Fax: 080-26723604
Email: anmolpublicationsbangalore@gmail.com

Concepts in Plant Physiology

First Edition, 2013

ISBN: 978-81-261-5377-0

PRINTED IN INDIA

Printed at Avantika Printers Pvt. Ltd., New Delhi

Contents

Preface

The field of plant physiology includes the study of all the internal activities of plants—those chemical and physical processes associated with life as they occur in plants. This includes study at many levels of scale of size and time. At the smallest scale are molecular interactions of photosynthesis and internal diffusion of water, minerals, and nutrients. At the largest scale are the processes of plant development, seasonality, dormancy, and reproductive control. Major subdisciplines of plant physiology include phytochemistry (the study of the biochemistry of plants) and phytopathology (the study of disease in plants). The scope of plant physiology as a discipline may be divided into several major areas of research. First, the study of phytochemistry (plant chemistry) is included within the domain of plant physiology. To function and survive, plants produce a wide array of chemical compounds not found in other organisms.

Photosynthesis requires a large array of pigments, enzymes, and other compounds to function. Because they cannot move, plants must also defend themselves chemically from herbivores, pathogens and competition from other plants. They do this by producing toxins and foul-tasting or smelling chemicals. Other compounds defend plants against disease, permit survival during drought, and prepare plants for dormancy. While other compounds are used to attract pollinators or herbivores to spread ripe seeds. Secondly, plant physiology includes the study of biological and chemical processes of individual plant cells. Plant cells have a number of features that distinguish them from cells of animals, and which lead to major differences in the way that plant life behaves and responds differently from animal life. For example, plant cells have a cell wall which restricts the shape of plant cells and thereby limits the flexibility and mobility of plants. Plant cells also contain chlorophyll, a chemical compound that interacts with light in a way that enables plants to manufacture their own nutrients rather than consuming other living things as animals do. Thirdly, plant physiology deals with interactions between cells, tissues, and organs within a plant. Different cells and tissues are physically and chemically specialized to perform different functions. Roots and rhizoids function to anchor the plant and acquire minerals in the soil. Leaves catch light in order to manufacture nutrients. For both of these organs to remain living, minerals that the roots acquire must be transported to the leaves, and the nutrients manufactured in the leaves must be transported to the roots. Plants have developed a number of ways to achieve this transport, such as vascular tissue, and the functioning of the various modes of transport is studied by plant physiologists. Fourthly, plant physiologists study the ways that plants control or regulate internal functions. Like animals, plants produce chemicals called hormones which are produced in one part of the plant to signal cells in another part of the plant to respond. Many flowering plants bloom at the appropriate time because of light-sensitive compounds that respond to the length of the night, a phenomenon known as photoperiodism.

The list of simple elements of which plants are primarily constructed—carbon, oxygen, hydrogen, calcium, phosphorus, etc.—is not different from similar lists for animals, fungi, or even bacteria. The fundamental atomic components of plants are the same as for all life; only the details of the way in which they are assembled differs. Despite this underlying similarity, plants produce a vast array of chemical compounds with unusual properties which they use to cope with their environment. Pigments are used by plants to absorb or detect light, and are extracted by humans for use in dyes. Other plant products may be used for the manufacture of commercially important rubber or biofuel. Perhaps the most celebrated compounds from plants are those with pharmacologicalactivity, such as salicylic acid (aspirin), morphine, and digitalis. Drug companies spend billions of dollars each year researching plant compounds for potential medicinal benefits.

This book explores innovative ideas in this subject.

—Editor

1

Photosynthesis: Carbon Reactions

How the Calvin Cycle Was Elucidated

The elucidation of the Calvin cycle was the result of a series of elegant experiments by Melvin Calvin and his colleagues in the 1950s, for which a Nobel prize was awarded in 1961. They used suspensions of the unicellular eukaryotic green alga *Chlorella* to trace the path of carbon. The Calvin cycle is also referred to as the reductive pentose phosphate cycle to distinguish the photosynthetic cycle from the oxidative pentose phosphate pathway, with which it shares several enzymes. Yet another designation of the Calvin cycle is photosynthetic carbon reduction cycle.

To elucidate the cycle, the investigators first exposed the algal cells to constant conditions of light and CO_2 to establish steady-state photosynthesis. Next they added radioactive $^{14}CO_2$ for a brief period to label the intermediates of the cycle. They then killed the cells and inactivated their enzymes by plunging the suspension into boiling alcohol. They separated the ^{14}C-labeled compounds from one another and identified them by their positions on two-dimensional paper chromatograms.

In this manner, 3-phosphoglycerate and the various sugar phosphates were identified as intermediates in carbon fixation. By exposing the cells to $^{14}CO_2$ for progressively shorter periods of time, Calvin's group was able to identify 3-phosphoglycerate as the first stable intermediate. Therefore, the other labelled sugar phosphates must be derived from a subsequent reduction of 3-phosphoglycerate.

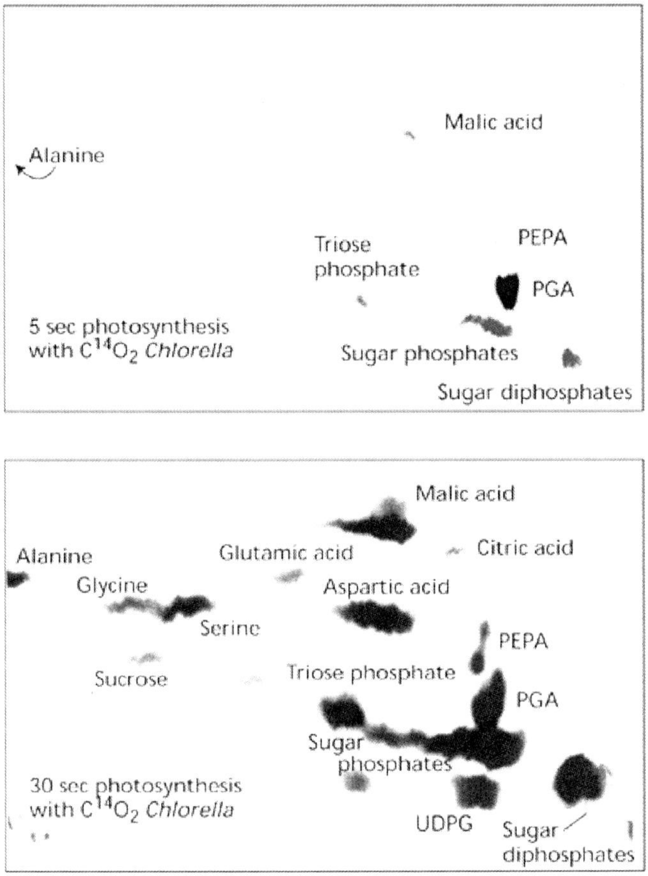

Figure: *Autoradiograms showing the labelling of carbon compounds in the alga Chlorella after exposure to $^{14}CO_2$. The time intervals shown in the figures indicate the length of exposure to the radiolabel. At the indicated time intervals, the reaction was terminated, by plunging of the contents into boiling alcohol. The labelled compounds in the cell homogenates were then separated by paper chromatography. The heavy labelling of 3-phosphoglycerate (PGA) after the shorter exposure indicates that it is the first stable intermediate of the Calvin (reductive pentose phosphate) cycle.*

To deduce the path of carbon, it was necessary to determine the distribution of ^{14}C in each of the labelled sugars. After a brief exposure to $^{14}CO_2$, individual intermediates were isolated and chemically degraded so that the amount of ^{14}C in each carbon atom could be determined. The results showed that 3-phosphoglycerate was initially labelled predominantly in the carboxyl group. This finding suggested that the initial CO_2 acceptor was a two-carbon compound, and it prompted a long and futile search for such a compound.

The subsequent discovery that pentose monophosphates and a pentose bisphosphate (ribulose) participate in the cycle raised the possibility that the initial CO_2 acceptor was a five-carbon compound. This conceptual breakthrough rapidly led to the identification of ribulose-1,5-bisphosphate as the CO_2 acceptor and to the formulation of the complete cycle. The operation of the cycle explained the ^{14}C-labeling pattern of the other intermediates.

To demonstrate conclusively that a metabolic pathway exists, it is necessary to prove that the postulated enzymes catalyze the proposed reactions in the test tube. Ideally, the rates of these enzyme reactions (in vitro) should be equal to or in excess of those observed in the intact cell (in vivo). However, this evidence can be used only to support a proposed pathway. Failure to demonstrate a reaction in vitro does not prove that the reaction does not occur.

All of the reactions of the Calvin cycle have been demonstrated in vitro. With one exception, the in vitro rates of enzymatic activity are well in excess of the maximal observed rates of photosynthesis. That exception is rubisco, which, when assayed under the levels of CO_2 and O_2 in air, shows barely enough in vitro activity to account for the observed rate of photosynthesis in air. Although great progress has been made in increasing activity by providing proper conditions for activation, researchers continue to debate whether rubisco is a rate-limiting step in photosynthesis.

Rubisco: A Model Enzyme for Studying Structure and Function

With the exception of the carbon fixed by a some prokaryotic organisms, most of the carbon fixed on Earth is processed by the Calvin cycle. The concentration of rubisco, the carboxylating enzyme of the Calvin cycle, is generally high. For example, it accounts for 50% or more of the total protein in plant leaves, and its concentration within chloroplasts is extremely high (*ca.* 0.2 g/ml).

Rubisco occurs in two functionally analogous forms, having different structure, distribution, and O_2 sensitivity. Oxygenic phototroph (cyanobacteria, chloroplasts from high plants) and many photosynthetic bacteria contain a form of the enzyme, (form I), made up of eight large (L) catalytic subunits (about 55 kDa each) and eight small (S) subunits (about 14 kDa each), giving a molecular mass of about 560 kDa for the complete protein (L_8S_8). The 8 L subunits of form I are arranged as an octameric core surrounded by two layers of four S subunits, with each layer located on opposite sides of the molecule.

Form II of rubisco is found in some photosynthetic purple nonsulfur bacteria, and it is composed of two L subunits, each 50 kDa. Eukaryotic algal dinoflagellates also use a dimeric enzyme of proteobacterial ancestry. A rubisco-like enzyme lacking catalytic activity has recently been identified in photosynthetic green sulfur bacteria which lack the Calvin cycle. The evidence suggests that the enzyme may function in sulfur oxidation to provide electrons for the photosynthetic electron transport chain. Genome sequences show that the rubisco-like protein also occurs in certain heterotrophic bacteria.

Rubisco Activase

Rubisco catalyzes the incorporation of CO_2 to the C-2 of the enediol form of ribulose-1,5-bisphosphate and the subsequent cleavage of the unstable intermediate to yield two molecules of 3-phosphoglycerate. The conversion of the inactive form of rubisco to an active state, which is required for both the carboxylation and oxygenation reactions, requires the carbamylation of a specific lysine residue (in spinach, Lys-201). As a consequence of the binding of the CO_2^- molecule to the lysine residue, the enzyme acquires a triad of anionic residues that provide the binding site for the cofactor Mg^{2+} (in spinach, the other two residues are Asp-203 and Glu-204). At this stage, the incorporation of ribulose-1,5-bisphosphate causes conformational changes in rubisco that allows the enzyme to sequester the sugar bisphosphate from the solvent and, upon the binding of a CO_2 or O_2 molecule, to start the catalytic cycle.

However, the tight binding of ribulose-1,5-bisphosphate to the uncarbamylated enzyme molecules displaces the equilibrium to the dead end complex, rubisco-ribulose-1,5-bisphosphate, so that the rate of reaction declines rapidly until it finally stops. How is this problem solved?

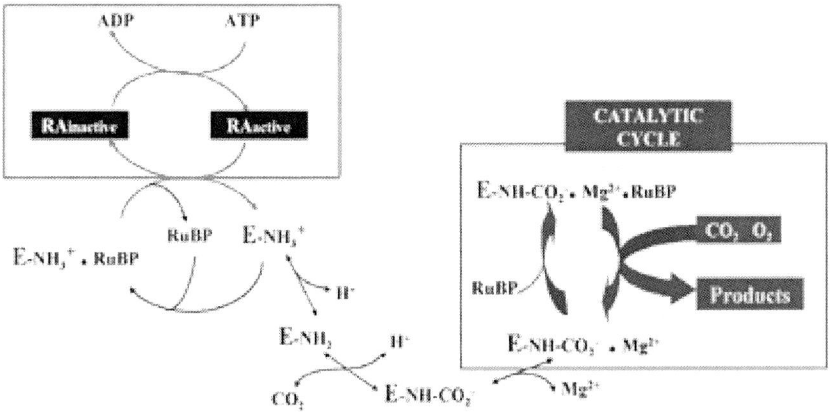

Figure: Modulation of rubisco by rubisco-activase. The combination or the interaction of rubisco (E) with modulators distributes the enzyme among different states (yellow boxes). The unprotonated form of a specific lysine residue (E-NH_2) reacts with CO_2 releasing a proton and forming the carbamate (E-NH-CO_2^-) that is stabilized by the cofactor Mg^{2+} (E-NH-CO_2^-. Mg^{2+}). At this stage (blue box), rubisco is catalytically competent and thereby binds ribulose-1,5-bisphosphate (RuBP) yielding the ternary complex (E-NH-CO_2^-. Mg^{2+}. RuBP) in which the latter is converted to the enediol form. This entails the incorporation of CO_2 or O_2 in the active site to generate the products of carboxylation or oxygenation, respectively. The high affinity of ribulose-1,5 bisphosphate for free rubisco (E-NH^{3+}) generates the dead-end complex (E-NH^{3+}. RuBP) thereby displacing the above equilibria to a catalytically incompetent state. Rubisco-activase (RA) interacts noncovalently with the binary complex and facilitates the release of ribulose-1,5-bisphosphate (red box) freeing the enzyme for the next carbamylation cycle. This process requires the concurrent hydrolysis of ATP.

A combination of genetic, physiological and biochemical studies have provided the solution to the long-term riddle of rubisco activation.

In vivo, some mutants of *Arabidopsis thaliana* lack light-mediated activation of rubisco and can only grow at high concentrations of CO_2. Surprisingly, the mutation does not impair the response of the enzyme to modulation by CO_2 and Mg^{2+} in vitro. These contradictory results led to the discovery of a protein, rubisco-activase, that uncouples ribulose-1,5-bisphosphate from decarbamylated active sites and, in so doing, promotes the access of CO_2 and Mg^{2+} for the carbamylation of the enzyme.

The importance of this process is that the rate at which rubisco-activase enhances the catalytic capacity of rubisco is linked to the hydrolysis of ATP (50 molecules of ATP hydrolyzed / 1 rubisco site activated).

Two isoforms of rubisco activase, (42 and 46 kDa) found in several species differ in the C-terminal region as result of the differential splicing of a common pre-mRNA. Both isoforms catalyze the activation of rubisco and the hydrolysis of ATP but the smaller isoform has higher capacity than the larger isoform for both activities.

Evidence for the fact that the two processes are independent are provided by studies showing that the digestion of native rubisco-activase with trypsin or the modification of its primary structure by site-directed mutagenesis do not elicit the same responses in the stimulation of rubisco and the hydrolysis of ATP. More importantly, the deletion of some residues at the C-region of rubisco-activase doubles the rate of rubisco activation with only a minor effect on the hydrolysis of ATP.

Rubisco activation by the activase requires a mutual recognition of both proteins. Rubisco-activase from tobacco (*Solanaceae*) fails to activate rubisco from spinach or *Chlamydomonas* (non-*Solanaceae*) and, similarly, the rubisco-activase from spinach or *Chlamydomonas* fails to stimulate rubisco from tobacco. However, the substitution of some residues on the surface of the large subunit of *Chlamydomonas* rubisco shifts the specificity from non-*Solanaceae* to *Solanaceae* rubisco-activase.

These experiments underscore the contribution of specific amino acids to protein-protein interaction and provide experimental evidence for a recognition step separated from the subsequent activation process.

Three Variations of C_4 Metabolism

There are three variations of the basic C_4 pathway. The variations differ principally in the C_4 acid transported into the bundle sheath cells (malate or aspartate) and in the manner of decarboxylation.

The variations are named after the enzymes that catalyze the respective decarboxylation reactions: NADP-dependent malic enzyme (NADP-ME) found in chloroplasts; NAD-dependent malic enzyme (NAD-ME) in mitochondria; and phosphoenolpyruvate carboxykinase (PEP-CK) in the cytosol. The discussion in the text traces the route of carbon for the NADP-ME type, such as sugarcane or corn.

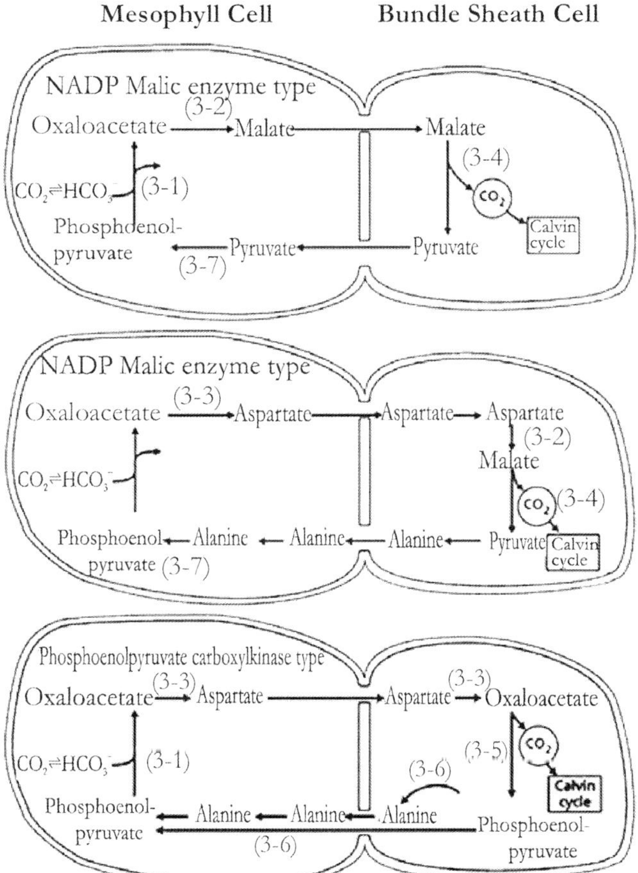

Figure: *The three variants of the* C_4 *photosynthetic carbon cycle. The variants differ principally in (1) the nature of the four-carbon acid (malate or aspartate) transported into the bundle sheath cell and of the three-carbon acid (pyruvate or alanine) returned to the mesophyll cell and (2) the nature of the enzyme that catalyzes the decarboxylation step in the bundle sheath cell. The three variants are named after the enzymes that catalyze the decarboxylation reactions. Representatives of each variant include maize, crabgrass, sugarcane, sorghum (NADP malic enzyme); pigweed, millet (NAD malic enzyme); guinea grass (phosphoenolpyruvate carboxykinase).*

The Carbon Path

The primary carboxylation reaction, catalyzed by PEP carboxylase, which is common to all three variants, occurs in the cytosol of the mesophyll cells and uses HCO_3^- rather than CO_2 as a substrate. In the NADP-ME variant, oxaloacetate is rapidly reduced to malate in the mesophyll chloroplasts by NADPH using NADP:malate dehydrogenase. The malate formed enters the chloroplast of the bundle sheath cell and there undergoes oxidative decarboxylation, yielding pyruvate. The CO_2 released within the bundle sheath cells is converted to carbohydrate by the Calvin cycle.

The residual C_3 acid is transported back to the mesophyll as pyruvate and converted to phosphoenolpyruvate in the mesophyll chloroplast. This reaction takes place in the mesophyll chloroplast of all three C_4 variants. The cycle then starts again with carboxylation of the phosphoenolpyruvate to produce oxaloacetate.

Another process localized in the mesophyll chloroplast of all three C_4 variants is the reduction of 3-phosphoglycerate to triose phosphates, which occurs within the Calvin cycle proper in C_3 chloroplasts. In contrast, all C_4 variants export 3-phosphoglycerate from the bundle sheath for reduction in the mesophyll chloroplast. After reduction, the triose phosphates are transported back to the bundle sheath chloroplasts, for use in the operation of the Calvin cycle there.

One possible reason for this remarkable split of the Calvin cycle between the bundle sheath and the mesophyll chloroplast is to restrict electron transport activity and oxygen evolution in the bundle sheath, and thus avoid oxygen levels that would favour photorespiration. The transport of metabolites of the C_4 pathway is accompanied by movements of protons and other ionic species in order to maintain pH and charge balance.

Photorespiration in CAM Plants

Many terrestrial plants, adapted to hot and arid regions where water is scarce, are stressed by the severe effects of full sunlight, particularly drought and high temperature. To avoid high rates of water transpiration, the stomata of these species are tightly closed during the hot diurnal hours and perform most of their gas exchange with the surrounding atmosphere at night, when the air is relatively cool and humid. On the other hand, the vital process of photosynthesis in which CO_2 is incorporated into carbohydrates takes place during the day.

To cope with the lack of an efficient diurnal influx of atmospheric CO_2, these species have evolved a different photosynthetic mechanism, called "crassulacean acid metabolism" (CAM). This unusual process takes place within single leaf cells in which the uptake and subsequent assimilation of atmospheric CO_2 involves three compartments: cytosol, vacuole, and chloroplast. In general, CAM plants are common among desert succulents (e.g., *Cactaceae*) exposed to daytime high temperatures and low relative humidity, or among tropical forest epiphytes (*Bromeliaceae*) which are exposed to physiological drought under humid conditions because of their aerial growth habit. Remarkably, CAM also occurs in aquatic species (*Isoetaceae*) found worldwide in marshy and shallow habitats that exhibit a diurnal limitation of CO_2.

During the cool night-time hours, the open stomata facilitate the uptake of atmospheric CO_2 which diffuses into the cell cytosol, generating HCO_3^-. At this stage, phosphoenolpyruvate-carboxylase (PEPCase) catalyzes the incorporation of HCO_3^- into phosphoenolpyruvate, yielding oxaloacetic acid and related four-carbon organic acids (particularly malic acid), that accumulate in the vacuole. During the hot hours of the day, the stomatal closure characteristic of terrestrial CAM plants reduces water loss but causes an internal CO_2 shortage. To overcome this deficiency, the vacuole releases malic acid into the cytosol where the action of decarboxylating enzymes, NADP–malic enzyme or PEP-carboxykinase, produces pyruvate and CO_2.

Inside the stroma of chloroplasts, Rubisco incorporates the released CO_2 into 3-phosphoglycerate which, in concert with the ATP and NADPH synthesized in the light via photophosphorylation, initiates a series of Benson–Calvin cycle reactions that result in the accumulation of starch. The temporal separation of the nocturnal formation and storage of malic acid, and the diurnal decarboxylation and the accompanying accumulation of plastid starch, proceeds over the course of a 24-hour photoperiod. On this temporal basis, CAM consists of four metabolic phases that encompass the modulation of C_4 (PEPCase) and C_3 (Rubisco) carboxylation within the same cellular environment (Figure 1).

The nocturnal uptake of atmospheric CO_2 via PEPCase to form C_4 acids and the diurnal decarboxylation of malic acid to elicit an elevated internal concentration of CO_2 are interspersed with transitional periods of net CO_2 uptake in early morning and in late afternoon, when both the PEPCase-and the Rubisco-mediated carboxylations contribute to the assimilation of CO_2. In this circadian rhythm, the proportion of CO_2 taken up via PEPCase at night or via Rubisco during the day is controlled by:

- stomatal function
- fluctuations in the content of organic acids and storage carbohydrates
- the activity of enzymes involved in the primary (PEPCase) and the secondary (Rubisco) carboxylations, and in the decarboxylation of organic acids (e.g., malic enzyme or PEP-carboxykinase)

In diurnal phase III, the closed stomata preclude gas exchange with the external atmosphere, but photosynthesis and other vital processes continue. At this stage, the internal levels of CO_2 fall precipitously due to an active Benson–Calvin cycle; however, this decline is offset by mitochondrial respiration.

The closure of the stomata also leads to the accumulation of O_2 in leaf cells because the rate of photosynthetic O_2 evolution is higher than mitochondrial respiration and other O_2-consuming reactions. At this stage, the oxygenase activity of Rubisco activates the photorespiratory cycle by which the Benson–Calvin cycle recovers 75% of the carbon diverted to glycolate, with the remaining 25% released as CO_2 by the mitochondrial glycine decarboxylase complex.

Thus, the closed stomata provide an adaptive advantage that minimizes water loss from the leaves by transpiration, and keeps the concentration of stromal CO_2 adequate for carbon assimilation. However, the concurrent accumulation of O_2 is a major adverse consequence because, as a powerful oxidant, O_2 reacts with other molecules and generates harmful free radicals (e.g., superoxide, hydroxyl ion) which may initiate uncontrolled adverse reactions on DNA, proteins, and lipids.

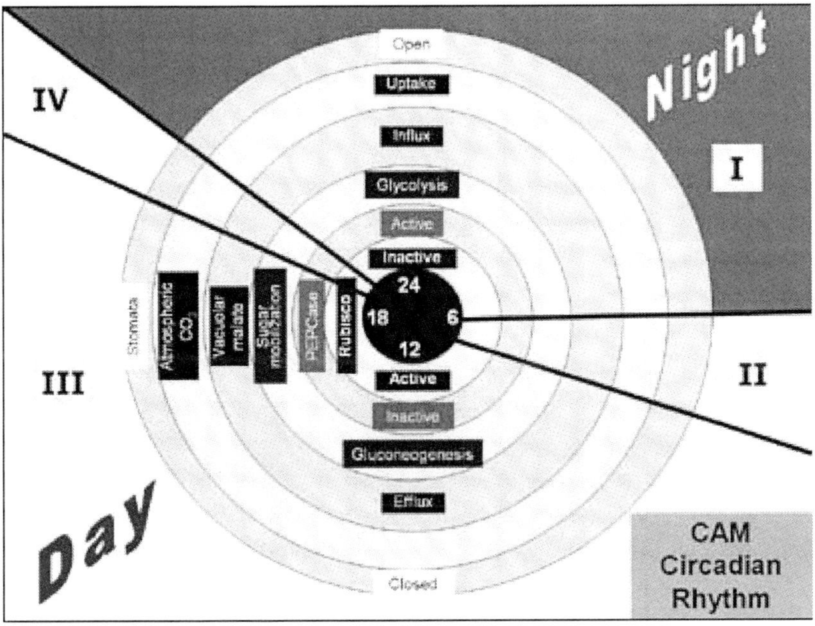

Figure: *CAM Circadian rhythm. Metabolite fluxes [atmospheric* CO_2*, vacuolar malate, sugar mobilization to form PEP] and enzyme activities [PEPCase, Rubisco] are indicated on the four major temporal phases of CAM [I, II, III, IV]. Phase I comprises the nocturnal PEPCase-driven uptake of atmospheric* CO_2 *and malic acid accumulation. Phase II is the transitional phase wherein an accelerated burst of* CO_2 *uptake takes place due to both the PEPCase-mediated and the Rubisco-mediated fixation of* CO_2*. During diurnal Phase III, the uptake of atmospheric* CO_2 *is diminished and the decarboxylation of the vacuolar malic acid generates the stromal concentration of* CO_2 *needed for Rubisco-mediated* CO_2 *fixation via the Benson–Calvin cycle. Phase IV may include* CO_2 *uptake that is fixed directly by Rubisco.*

In this scenario, photorespiration helps alleviate the undesirable consequences of O_2 accumulation by consuming both stromal and peroxisomal O_2. The release of one mole CO_2 in a single round of the photorespiration cycle requires the successive utilization of: (a) two moles O_2 by the oxygenase activity of Rubisco in the chloroplast stroma, and (b) one mole O_2 for the concerted action of glycolate oxidase and catalase in the small oxygen cycle in the peroxisome. The photorespiratory cycle driven by ribulose-

1,5-bisphosphate {2 Ribulose-1, 5-bisphosphate + 3 O_2 + H_2O + ATP '! 3 3-phosphoglycerate + CO_2 + 2 P_i + ADP} requires substantially more O_2 than mitochondrial respiration operating with carbohydrates {$C_6H_{12}O_6$ (e.g., glucose) + 6 O_2 '! 6 CO_2 + 6 H_2O}. Thus, three versus one moles O_2 are consumed, respectively, per mole CO_2 released. It is therefore evident that the beneficial consequence of CAM photorespiration in Phase III involves not only the avoidance of carbon loss due to the formation of glycolate, but also the amelioration of oxidative stress due to removal of reactive oxygen species. This mitigation of oxidative stress has been proposed to be a major driving force in the evolution of CAM.

Fructans

Although starch is the dominant carbohydrate reserve in many species, it may be almost absent in the storage organs of some plants. Fructans are less familiar reserve carbohydrates but occur in 15% of flowering species that mainly belong to the Asteraceae, Campanulaceae and Boraginaceae (dicots), and Poaceae and Liliaceae (monocots).

This type of polysaccharide, whose basic unit is fructose, was found in 1804 in a hot water extract from *Inula helenium* and later named "inulin." Fructans can replace starch as a reserve carbohydrate in some plants, though in many species both polysaccharides coexist. Moreover, the carbohydrate reserve substances vary among plant species and even within a single plant species throughout the seasons and under various climatic conditions. In contrast to starch—which is stored in plastids—fructans appear in vacuoles of different organs: for example, in bulbs (onion), tubers (Jerusalem artichoke, dahlia), taproots (chicory), leaves, and in stems (wheat, barley).

Chemically, fructans are distinguished on the basis of the glycosidic bond that links fructose residues to each other. The biosynthesis of these polysaccharides starts with the incorporation of a fructose moeity to one of the three primary hydroxyl groups of vacuolar sucrose—that is, carbon 1 and carbon-6 in the fructose moeity, and carbon-6 in the glucose moeity (Figure above). This reaction, catalyzed by the enzyme sucrose:sucrose 1-fructosyl transferase, implies a transglycosylation in which one molecule of sucrose transfers the fructosyl moeity to another.

Thus, the addition of β-D-2,1-or β-D-2,6-linked fructofuranosyl units to the fructose moeity of sucrose yields 1-kestose and 6-kestose, respectively, while the addition of a β-D-2,6-linked fructofuranosyl unit to the glucose moeity of sucrose produces neo-kestose.

Further elongation of the ketose-type trisaccharide proceeds by the action of specific enzymes that transfer, also via transglycosylation, a fructosyl unit from a fructan molecule to positions 1 or 6 in the second fructose moeity of the trisaccharide. Thus, fructan:fructan fructosyl transferases catalyze the elongation of kestoses yielding the linear β-D-2,1-bonds of inulins, mainly in dicots, and the β-D-2,6-linkages in levans of monocots and bacteria.

$$\text{GFF (kestose)} + G(F)_m \rightarrow G(F)_3 + G(F)_{m-1}$$

While up to 100,000 fructose moeities can be linked in a single molecule of bacterial fructan, fructosyl units range from 10 to 200 in plant counterparts. Fructans generally are present as a continuum of oligofructans that differ from each other in one fructose residue: $G(F)_3$, $G(F)_4$, $G(F)_5$, $G(F)_6$,..., $G(F)_n$. This feature enables fructans to participate in the regulation of the cellular osmotic potential.

In addition to supplying hexose units for demands of cellular energy, the breakdown of the endogenous polysaccharide elevates the concentration of lower members in the oligofructans series, quickly increasing in consequence the osmotic pressure.

Two types of fructan hydrolases cleave the glycosidic bond in fructans: exohydrolases that release terminal fructosyl residues while endohydrolases cleave at random within the fructan chains. Apparently, plants contain only the former type while bacteria and fungi contain both kinds of enzymes.

The recent finding of fructan exohydrolases in nonfructan plants not only suggests that these enzymes are additionally involved in other important functions—perhaps defence and signalling—but also highlights the necessity of functional characterization of these enzymes (Van den Ende et al. 2004).

Plastidic Phosphate Translocators

Chloroplasts are metabolic powerhouses within leaf cells. They are not only the main site of carbon, nitrogen, and sulfur assimilation but also share metabolic pathways—such as the photorespiratory carbon cycle—with other plant cell compartments.

A continuous exchange of metabolites and ions with the cytosol is necessary for playing these roles.

In all vascular plants, plastids are surrounded by an envelope that restricts the nonspecific diffusion of polar molecules. Thus, a pair of concentric membranes modulate the exchange of metabolites between the stromal interior and the cytosol. Although many pore-forming proteins that are substrate-specific were characterized in the outer membrane, the inner membrane is considered the main permeability barrier between the cytosol and the chloroplast. All transporters of the inner membrane are nuclear-encoded membrane proteins whose precursors are synthesized in the cytosol and post-translationally imported into the inner membrane.

Chloroplast transporters collectively grouped as phosphate translocators catalyze a strict 1:1 counter-exchange (antiport) of orthophosphate with triose phosphates, pentose phosphates, and hexose phosphates. The Arabidopsis genome contains six genes that encode four classes of functional plastidial phosphate translocators (Table 1): single copies of the triose phosphate/phosphate translocator (TPT) and the phosphoenolpyruvate/phosphate translocators (PPT); and two copies each of the glucose 6-phosphate/phosphate translocators (GPT) and the xylulose 5-phosphate/phosphate translocator (XPT).

Table: *Plastidic Phosphate Translocators.*

Translocator	*MATDB entry*	*Compounds transported*
TPT	TPT (At5g46110) phosphoglycerate	triose phosphates, 3
PPT	PPT1 (At5g54800), PPT2 (At3g01550)	phosphoenolpyruvate, 2-phosphoglycerate
GPT	GPT1 (At5g33320), 6-phosphate, triose phosphates	GPT2 (At1g61800) glucose
XPT	(At5g17630) phosphates	xylulose 5-phosphate, triose

The Triose Phosphate/Phosphate Translocator and the Allocation of Carbon in Plants

Triose phosphates, generated by the Calvin cycle at the expense of photosynthetically generated ATP and NADPH, flow to the cytosol across the chloroplast envelope.

In turn, the orthophosphate released in the cytosol from biosynthetic processes is transported back into chloroplasts to replenish the ATP necessary for sustaining the assimilation of CO_2 and photosynthetic electron transport.

Both the efflux of triose phosphates and the influx of orthophosphate are driven by a dimer composed of two identical subunits—the chloroplastic TPT. This membrane protein functions as an antiport system that exchanges phosphorylated—mainly three-, but also admits five-or six-carbon compounds—for orthophosphate.

Hence, the chloroplast TPT exports the fixed carbon for the cytosolic synthesis of sucrose, the main product allocated to heterotrophic plant organs for further metabolism or conversion into storage products (e.g., starch, fructans).

Therefore, the chloroplast TPT represents *the diurnal path for carbon export.*

Mutants defective in chloroplast TPT rely on different paths to transfer the photoassimilate from the chloroplast to the cytosol and, in so doing, compensate for the deficiency in TPT activity.

Therefore, knockout mutations of this translocator increase the allocation of recently fixed carbon into transitory starch, followed by the degradation of the starch in the light and the subsequent export of neutral sugars to the cytosol.

On the other hand, the simultaneous knockout of chloroplast TPT and starch synthesis drastically impairs plant growth.

The Phosphoenolpyruvate/Phosphate Translocator is Essential for the Chloroplast Metabolism

PPT in the inner chloroplast membrane also links the stromal metabolism of all plants with the surrounding cytosol. C_4-plants of the NADP-malic enzyme type (e.g., maize) under active photosynthesis produce pyruvate in bundle sheath cells.

This three-carbon compound returns to mesophyll cells for the conversion to phosphoenolpyruvate by the chloroplast-localized pyruvate-phosphate dikinase (Figure below). Subsequently, the phosphoenolpyruvate flows from the stroma to the cytosol for the primary carbon fixation via PEPCase. To ensure high rates of phosphoenolpyruvate efflux for the proper functioning of this cycle, the envelope of mesophyll chloroplasts bears the PPT, which concurrently drives the chloroplast uptake of orthophosphate to restore the lost phosphorus.

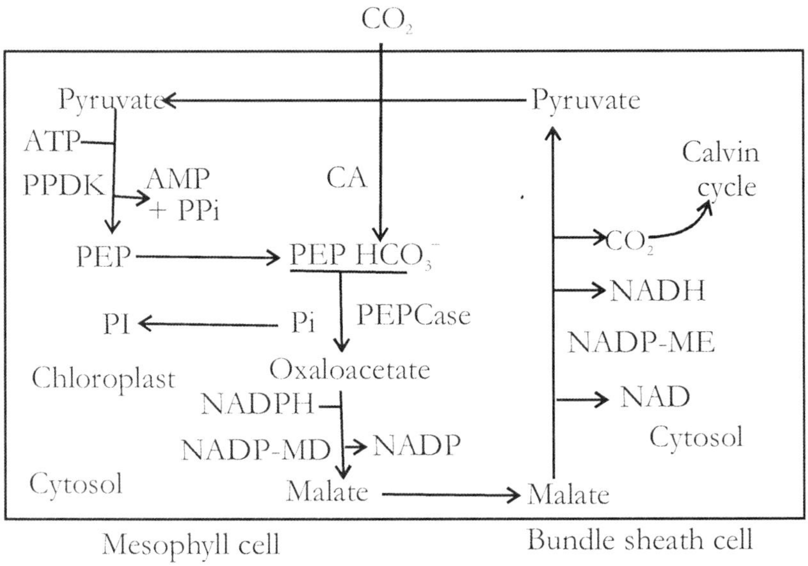

Figure: *Mobilization of phosphoenolpyruvate in mesophyll chloroplasts of C_4 plants of the NAD-malic enzyme type. The pyruvate produced by the malic enzyme (ME) in bundle sheath cells flows to mesophyll cells for conversion to phosphoenolpyruvate (PEP) by the pyruvate-phosphate dikinase (PPDK) localized at the chloroplast stroma. The efflux of PEP to the cytosol through the phosphoenolpyruvate/phosphate translocator (PPT) restores the acceptor of HCO_{3-} required by the phosphoenolpyruvate carboxylase (PEPCase) for the primary carbon fixation. CA: carbonic anhydrase; NADP-MD: NADP-dependent malate dehydrogenase.*

On the other hand, CAM plants of the malic enzyme-type decarboxylate during the day the malate accumulated at night. Again, two successive processes account for the production of phosphoenolpyruvate, the primary acceptor of CO_2. First, the pyruvate formed in decarboxylation of the malate enters into the chloroplast stroma for its transformation to phosphoenol-pyruvate catalyzed by the pyruvate-phosphate dikinase. Second, the phosphoenolpyruvate leaves the chloroplast in exchange for orthophosphate via the PPT.

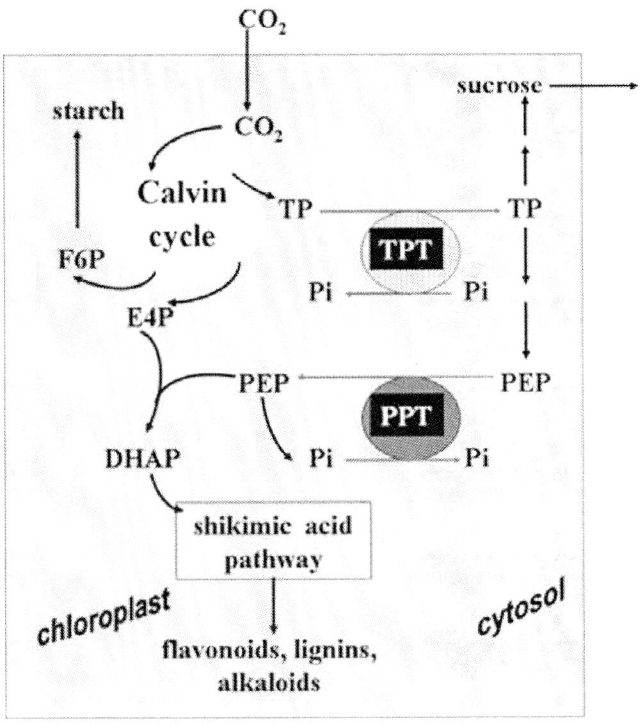

Figure: *Mobilization of phosphoenolpyruvate in chloroplasts of C_3 plants. Under active photosynthesis, TPT furnish the cytosol with carbon skeletons for the biosynthesis of phosphoenolpyruvate. Upon the release of orthophosphate, the uptake of the cytosolic phosphoenolpyruvate, which combines in the stroma with erythrose 4-phosphate, triggers the formation of 3-deoxy-D-arabino-heptulosonic acid-7-phosphate (DHAP). DHAP is a scaffold for the biosynthesis of shikimic acid-5-phosphate, which in turn starts the pathway leading to the aromatic amino acids tyrosine and phenylalanine, the precursors of flavonoids, lignins, and alkaloids.*

Knowledge on the function of the PPT in C_3 plants or nongreen tissues of C_4 plants is scarce. Apparently, the uptake—rather than the export—of phosphoenolpyruvate is the primary function of PPT in these tissues. Chloroplasts and most nongreen plastids are devoid of phosphoglyceromutase and enolase and, consequently, are unable to convert 3-phosphoglycerate to phosphoenolpyruvate. However, phosphoenolpyruvate combines with erythrose 4-phosphate, an intermediate of the pentose-phosphate pathway, for the synthesis of aromatic compounds through the shikimic acid pathway (Figure below). Thus, the provision of cytosolic phosphoenolpyruvate to the stroma via the PPT would be essential for the biosynthesis of flavonoids, lignins, alkaloids.

The Glucose 6-Phosphate/Phosphate Translocator Functions in Nongreen Tissues and is Essential for Plant Development

The genome of Arabidopsis harbours two paralogous GPT genes, *AtGPT1* and *AtGPT2*, that, apparently, function in plastids of nongreen tissues for the import of glucose 6-phosphate. Two transgenic lines, bearing T-DNA insertions in the *GPT1* gene, exhibited not only impaired pollen and ovule development but also were lethal in the homozygous state.

At variance, the disruption of the *GPT2* genes have no effect on growth and development. On the basis of drastic gametogenesis defects, it appears that the import of glucose 6-phosphate into nongreen plastids by the translocator AtGPT1 is crucial for the maturation of the pollen and the development of the female gametophyte.

The Xylulose 5-Phosphate/Phosphate Translocator

The XPT constitutes the fourth subfamily of phosphate translocators that exhibits substrate specificity similar to the GPT but lacks the capacity to transport glucose 6-phosphate. In Arabidopsis, the lack of cytosolic transketolase and transaldolase halts the cytosolic oxidative pentose phosphate pathway at the stage of pentose phosphates (Kruger and von Schaewen 2003). In this context, the putative function of the XPT would be to furnish the stromal pentose phosphate pathway with cytosolic five-carbon skeletons in the form of xylulose 5-phosphate, especially under conditions that require its removal from the cytosol.

Photosynthesis: Physiological and Ecological Considerations

Working with Light

Three light properties are especially important when working with light: amount, direction, and spectral quality. The first two parameters, amount and direction, are important with respect to the geometry of the part of the plant that intercepts the light. Is the plant part flat or cylindrical? For a flat leaf, a planar light sensor is the most appropriate, and the amount of energy that falls on a flat sensor of known area per unit time is quantified as irradiance. Units can be expressed in terms of energy, such as watts per square meter (W m^{-2}). Time (seconds) is contained within the term watt: 1 W = 1 joule (J)s^{-1}. The energy of a photon depends on its frequency, as expressed by Planck's law.

When considered as a wave, light has a wavelength and a frequency. Light can also be thought of as a stream of particles, photons, or quanta. In this case, units can be expressed in moles per square meter per second (mol m^{-2} s^{-1}), where "moles" refers to the number of photons (1 mol of light = 6.02 × 10^{23} photons, Avogadro's number). This measure is called photon irradiance.

Quanta and energy units can be interconverted, provided that the wavelength of the light is known. The energy of a photon is related to its wavelength as follows:

$$E = \frac{hc}{\lambda}$$

where *c* is the speed of light (3 × 10^8 m s^{-1}), *h* is Planck's constant (6.63 × 10^{-34} J s), and λ is the wavelength of light, usually expressed in nm (1 nm = 10^{-9} m). We can solve for the *hë* part of the equation, and we obtain 1,988 × 10^{-16}, and write this equation as:

$$E = \frac{1{,}988 \times 10^{-16}\,J}{\lambda}$$

where λ is expressed in nanometers. From this equation we can see that a photon at 400 nm, which is in the blue region of the spectrum, has twice the energy of a photon at 800 nm, from the infrared region of the spectrum. A photon of 400 nm light contains 4.97 × 10^{-19} J. On the other hand, the 800 nm photon contains 2.48 × 10^{-19} J. Stated differently, the higher the wavelength of a photon, the lower its energy, as indicated by the larger denominator in the equation.

Now, suppose we have 3 μmol of 400 nm light falling on 1 m^2 every second, or a photon irradiance of 3 μmol m^{-2} s^{-1}. This quantity is approximately the amount of blue light at

400 nm that strikes the surface of Earth on a sunny day. If we want to convert photon irradiance to energy irradiance, we must first convert micromoles to moles (1 μmol = 10^{-6} mol): (3 μmol m^{-2} s^{-1}) × (10^{-6} mol μmol^{-1}) = 3 × 10^{-6} mol m^{-2} s^{-1} of photons, or quanta. Calculating the number of quanta from Avogadro's number gives: (3 × 10^{-6} mol quanta m^{-2} s^{-1}) × (6.02 × 10^{23} quanta mol^{-1}) = 1.8 × 10^{18} quanta m^{-2} s^{-1}. As we already calculated, each photon, or quantum, of 400 nm light contains 4.97 × 10^{-19} J. Thus (4.97 × 10^{-19} J $quantum^{-1}$) × (1.8 × 10^{18} quanta m^{-2} s^{-1}) = 0.9 J s^{-1} m^{-2}. Since 1 J s^{-1} = 1 W, we have an irradiance of 0.9 W m^{-2}.

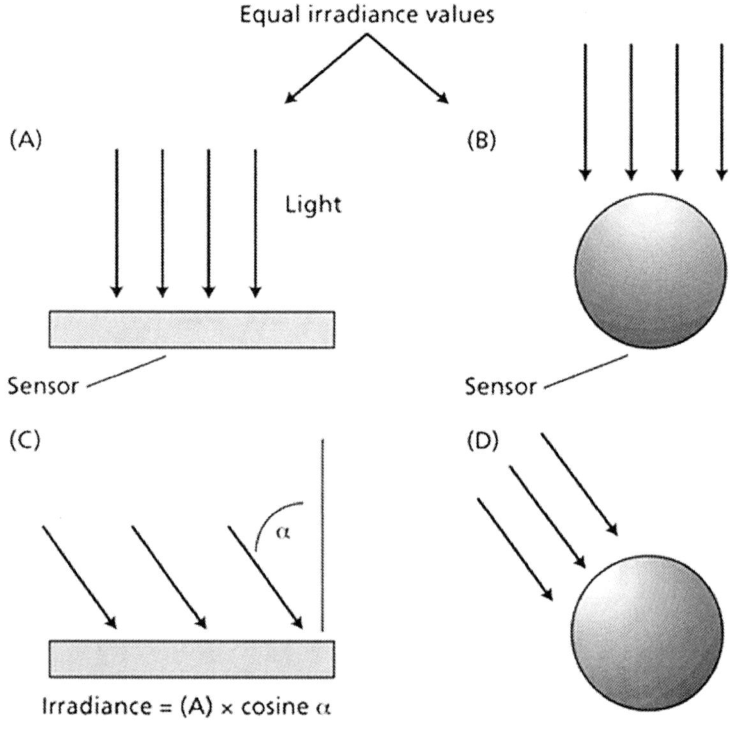

***Figure:** Irradiance and fluence rate. Equivalent amounts of collimated light strike a flat irradiance-type sensor (A) and a spherical sensor (B) that measure fluence rate. With collimated light, A and B will give the same light readings. When the light direction is changed 45°, the spherical sensor (D) will measure the same quantity as in B. In contrast, the flat irradiance sensor (C) will measure an amount equivalent to the irradiance in A multiplied by cosine of the angle α in C. (After Björn and Vogelmann 1994.)*

Direction of light. Turning our attention to the direction of light, light can strike a flat surface directly from above or it can strike the surface obliquely. When light deviates from perpendicular, irradiance is proportional to the cosine of the angle at which the light rays hit the sensor.

Thus, irradiance is maximal when light strikes a surface directly from above, and it decreases as light becomes more oblique—similar to the situation with a typical leaf. Sensors that correct for the angle of incidence of light are said to be *cosine corrected.*

There are many examples in nature in which the light-intercepting object is not flat (e.g., complex shoots, whole plants, chloroplasts). In addition, in some situations light can come from many directions simultaneously (e.g., direct light from the sun plus the light that is reflected upward from sand, soil, or snow). In these situations it makes more sense to measure light with a spherical sensor that measures light omnidirectionally (from all directions).

When the amount of light is measured by this omnidirectional measurement, the type of measurement is called fluence rate (Rupert and Letarjet 1978), and the measured amount of light can be expressed in watts per square meter ($W\ m^{-2}$) or moles per square meter per second ($mol\ m^{-2}\ s^{-1}$). It is clear from the units whether light is being measured as energy (W) or as photons (mol).

In contrast to a flat sensor, a spherical sensor is equally sensitive to light from all directions. Depending on whether the light is collimated (rays are parallel) or diffuse (rays travel in random directions), values for fluence rate versus irradiance can be quite different from one another. They are equivalent only under special conditions. The flat sensor measurement of photosynthetically active radiation (PAR, 400 to 700 nm) may also be expressed on the basis of energy ($W\ m^{-2}$) or quanta ($mol\ m^{-2}\ s^{-1}$) (McCree 1981). It is important to note that PAR is an irradiance-type measurement. In research on photosynthesis, when PAR is expressed on a quantum basis, it is often given the special term photosynthetic photon flux density (PPFD). However, it has been suggested that the term "density" be discontinued because within the International

System of Units (Système Internationale d'Unités, or SI units) "density" can mean area or volume. Moreover, area is contained within the term flux. PPFD has in some cases been shortened to PPF, but it is not clear whether this abbreviation represents an irradiance-type or a spherical measurement.

When light moves through water such as lakes or oceans, it scatters in such a way that its *direction* of travel is randomized. As a result, aquatic photosynthetic organisms can receive light simultaneously from above, below, and the sides. In oceanography and limnology, the light that travels from the water surface toward the bottom is called *downwelling irradiance*, and the light that travels in the reverse direction is called *upwelling irradiance*.

Before the advent of specialized light meters suitable for plant physiology, light was measured in lux or foot-candles. These measurements are based on the perception of light by the human eye, which is maximally sensitive to light within the green region of the spectrum, at 555 nm. Sensitivity falls off on both sides of this wavelength and approaches zero within the blue and red, wavelength regions that are important for photosynthesis.

Thus, lux and foot-candles are of little use in plant physiology. These units can be converted to the units of irradiance and fluence rate, but only if detailed knowledge about the spectral quality of the light is available, something that can be obtained only by use of an instrument called a *spectroradiometer*, which measures the amount of light at each wavelength.

In summary, when choosing how to quantify light, it is important to match sensor geometry and spectral response with that of the plant. Flat, cosine-corrected sensors are ideally suited to measure the amount of light that strikes the surface of a leaf; spherical sensors are more appropriate in other situations, such as when studying a chloroplast suspension or a branch from a tree.

How much light is there on a sunny day and what is the relationship between PAR irradiance and PAR fluence rate? Under direct sunlight, PAR irradiance and fluence rate are both about 2000 μmol m^{-2} s^{-1}, though higher values can be measured at high altitudes. The corresponding value in energy units is about 400 W m^{-2}. When light is completely diffuse, irradiance is only 0.25 times the fluence rate.

Heat Dissipation from Leaves: The Bowen Ratio

The heat load on a leaf exposed to full sunlight is very high. In fact, a leaf with an effective thickness of water of 300 μm would warm up by 100°C every minute if all available solar energy were absorbed and no heat was lost. However, this enormous heat load is dissipated by the emission of long-wave radiation, by sensible (or perceptible) heat loss, and by evaporative (or latent) heat loss.

Air circulation around the leaf removes heat from the leaf surfaces if the temperature of the leaf is higher than that of the air; this phenomenon is called sensible heat loss.

Evaporative heat loss occurs because the evaporation of water requires energy. Thus, as water evaporates from a leaf it withdraws heat from the leaf and cools it. The human body is cooled by the same principle, through perspiration.

Sensible heat loss and evaporative heat loss are the most important processes in the regulation of leaf temperature, and the ratio of the two is called the Bowen ratio:

$$\text{Bowen ratio} = \frac{\text{sensible heat loss}}{\text{evaporative heat loss}}$$

This concept was developed by Ira S. Bowen (1898–1978), an American astrophysicist. When the evaporation rate is low, because water supply is limited, the Bowen ratio tends to be high. Thus, the Bowen ratio is about 10 for deserts, 2-6 for semi-arid regions, 0.4 to 0.8 for temperate forests and grasslands, 0.2 for tropical rain forests and 0.1 for tropical oceans.

In well-watered crops, transpiration, and hence water evaporation from the leaf, is high, so the Bowen ratio is low. On the other hand, in some cacti, stomata closure prevents evaporative cooling; all the heat is dissipated by sensible heat loss, and the Bowen ratio is infinite. Plants with very high Bowen ratios conserve water but have to endure very high leaf temperatures in order to maintain a sufficient temperature gradient between the leaf and the air. Slow growth is usually correlated with these adaptations.

Very low Bowen ratios can be measured in a lawn on a relatively still day. In these conditions there is no sensible heat loss, because the air around the leaf is at the same temperature as the leaf; the Bowen ratio therefore approaches zero, and heat dissipation is due mostly to evaporative heat loss.

In other cases, such as cotton leaves in the afternoon, water loss from stomatal transpiration cools the leaf below the air temperature by evaporative heat loss. In that case, there is sensible heat gain rather than heat loss, and the Bowen ratio becomes negative.

One can calculate the evapotranspiration rate for an entire canopy using measurements of the Bowen ratio, net incident radiation, the heat loss from the soil, and the gradients in temperature and water vapour concentration above the canopy.

The Geographic Distributions of C_3 and C_4 Plants

Among the 15,000+ species with C_4 photosynthesis, it is most common in grasses and sedges, less common in herbs and shrubs, and not found in trees (with a single Hawaiian tree exception, *Euphorbia forbesii*). Climate is a major factor influencing the natural distributions of C_3 plants and C_4 grasses.

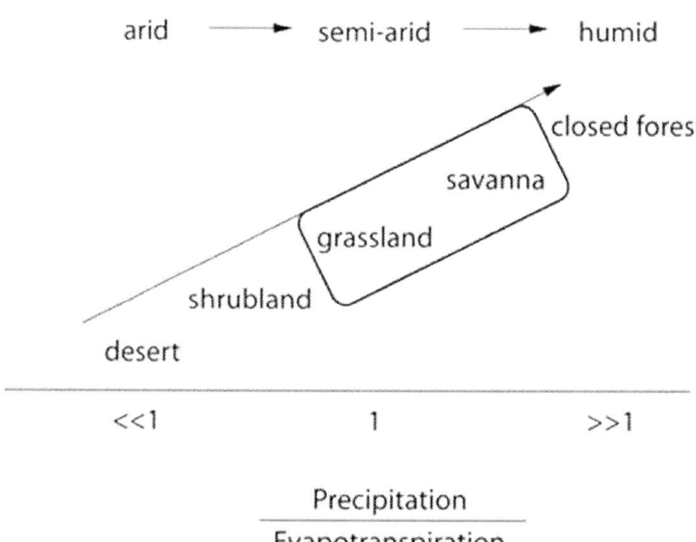

Figure: A plot of the vegetation types that are expected to occur along a climate gradient estimated by the ratio of the incoming precipitation to the potential evaporation rate. Courtesy of Ehleringer, unpublished figure.

Here, we talk about the two most important climate parameters influencing plant growth: water and temperature. Clearly plants will not grow in the absence of water, so the important factor influencing photosynthetic pathway distribution becomes the temperature during the growing period. Based on many systematic surveys of the natural vegetation across the globe that have been accumulating over time, a clear picture is emerging. C_4 taxa are found in warm to temperate environments and are uncommon in cool to cold climates. Below we construct a global map that describes the general abundances of C_3 and C_4 taxa on different continents.

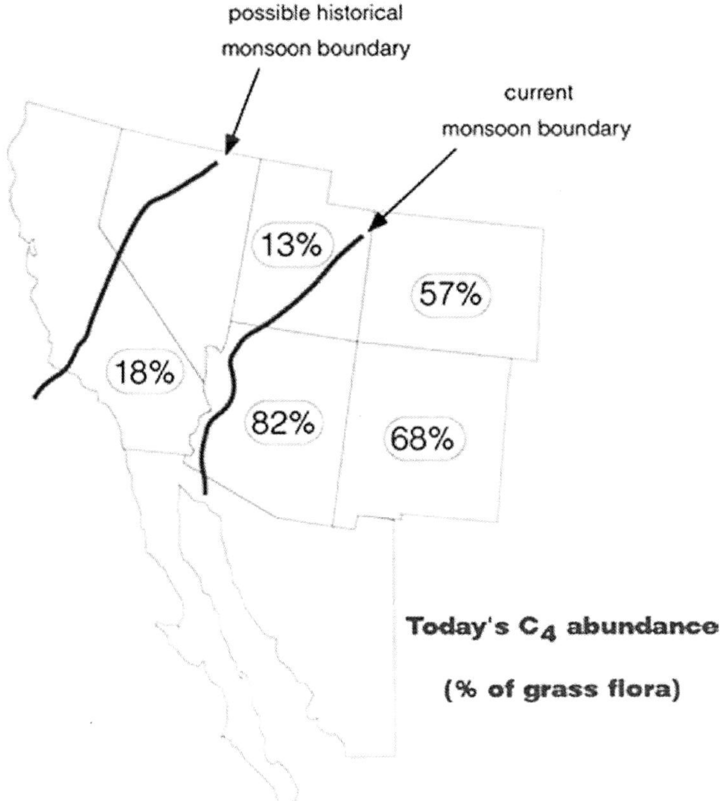

Figure: *Calculations of the percentage of the grass species that have C_4photosynthesis in different states of the western United States. Note that the presences of summer rain influences C_4 abundance. C_4 plant tend to occur in warm, summer wet regions. Courtesty of Ehleringer, Cerling, and Dearing (2005).*

Notice that C_4 taxa are not very common in tropical regions (0–20° latitude), because dense tropical forests tend to shade out

C_4 grasses. The C_4 taxa are most common away from the tropics in the savanna and steppe regions; their abundances tends to diminish south of the desert zones (generally 30–40° latitude). Climate can be presented along an axis of an increasing of precipitation relative to evaporative water loss. At the upper, wetter end of this gradient, we would find the forests. At the drier end, where precipitation is low relative to evaporative demand, we see the deserts. In the figure below, we see that these different vegetation zones reflect variations in the ratios of precipitation-to-evaporation. C_4 plants are most common in this semi-arid region.

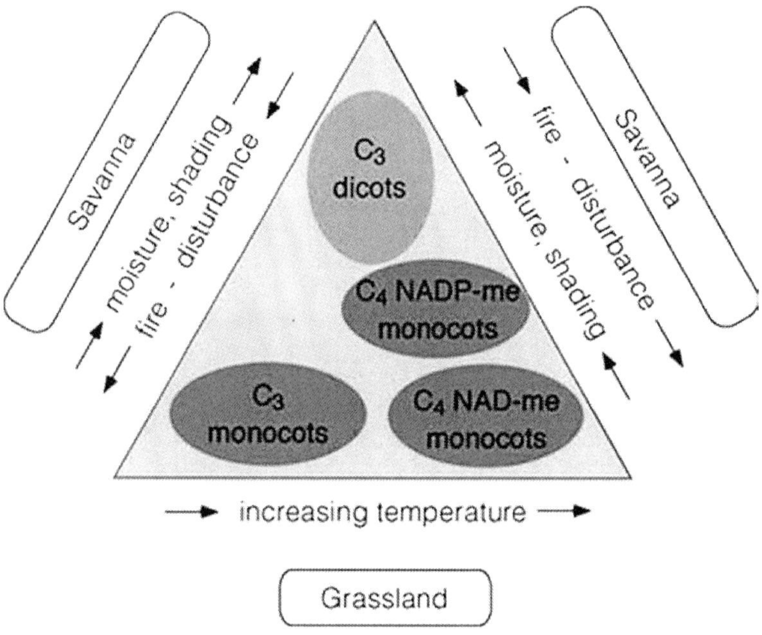

***Figure:** A three-axis, triangular presentation of the different gradients influencing the abundances of different C_4 photosynthesis subtypes. Courtesy of Ehleringer, Cerling, and Dearing (2005).*

In deserts with wet monsoon summers, such as in central Africa, the sub-Indian continent, and western North America, C_4 taxa are abundant. In deserts that receive winter rainfall, such as deserts in Asia and North Africa, there are few C_4 taxa. In North America, this regional influence of summer precipitation is most visible in the southwestern United States where there are large differences in precipitation associated with the Arizona monsoon boundary as shown below.

Human activities and disturbances by animals will influence the distribution of C_3 and C_4 taxa within savannas and grasslands. This reflects the role of disturbances, such as grazing and fires, on the importance of trees on the landscape.

Across the grasslands and savannas, the two dominant C_4-photosynthesis subtypes do not share identical distributions. The C_4 NADP-me grasses tend to occur in drier regions, such as the shortgrass prairie of the Great Plains. On the eastern, wetter edge of the Great Plains, the shortgrass prairie is replaced by tallgrass prairie, dominated by C_4 NAD-me grasses.

Today, agricultural practices result in C_4 plants growing outside of the distributions shown above. This results from the extensive planting of corn, sugarcane, and millet in both tropical and temperate regions.

2

Plant Cells

The Plant Kingdom

Since the time of Aristotle (384–322 B.C.E.), biologists have sought to classify organisms. At first the purpose was ease of identification ("artificial" classification schemes). Carolus Linnaeus (1707–1778), arguably the greatest of the pre-modern Naturalists, sought to classify plants and other organisms according to affinity groups that reflected the mind of the Creator. Later, after Darwin, the goal of classification was to show evolutionary relationships ("natural" classification schemes).

For the past 150 years, biologists have emphasized natural systems of classification and have attempted to define morphological criteria that reveal evolutionary relationships.

We now know that morphology, the form and structure of organisms, is the end product of the actions of genes. Virtually all of the information needed to form a complete organism is encoded in its DNA sequences, both nuclear and cytoplasmic (mitochondria and chloroplasts).

DNA sequence analysis has thus provided evolutionary biologists with a powerful new tool for arriving at a truly natural classification system.

On the basis of phylogenetic analyses of highly conserved DNA sequences, living organisms have been divided into three major domains: Bacteria, Archaea, and Eucarya.

1. The common ancestor of all the organisms first gave rise to the Bacteria and the common ancestor of the Archaea and the Eucarya.

2. The Archaea branch off from the Eucarya lineage.
3. The Eucarya common ancestor acquires mitochondrial endosymbiont.
4. A heterogeneous group of eukaryotes called protists branch off the lineage leading to plants, fungi, and animals.
5. The common ancestor of fungi and animals form a branch, followed by a divergence into the fungal and animal lineages.
6. The common ancestor of plants, green algae, red algae, and glaucophytes acquires chloroplast endosymbiont (a cyanobacterium).
7. The three lineages of glaucophytes, red algae, and green algae diverge.
8. Various lineages of protists acquire chloroplasts via green or red algal endosymbionts.
9. The earliest branch of green algae diverge.
10. The later branch of green algae diverge.
11. The remaining lineage leads to plants.

The Eucarya include the eukaryotes, organisms whose cells contain a true nucleus. The Bacteria, or eubacteria, which include the cyanobacteria, lack a true nucleus and are therefore prokaryotic.

The Archaea, or archaebacteria, are also prokaryotic, but they differ from the Bacteria: Besides their morphological and biochemical differences, they are often adapted to extreme environments, such as sulfur hot springs or saline ponds. Phylogenetic studies have indicated that the Archaea and Eucarya split after the Bacteria separated from the common ancestor. Thus Archaea and Eucarya represent sister groups. This closer relation between Archaea and Eucarya is reflected in their similar promoter structures and RNA polymerases, the presence of histones, and many other characteristics.

Fungi were formerly classified as algal-like plants that had lost their chloroplasts. However, fungi and animals branched off from the Eucarya lineage *before* the appearance of choloroplasts. They are thus more closely related to animals than to plants. Fungi are heterotrophic; that is, they depend on other organisms for their food, and they satisfy their nutritional needs by absorbing inorganic ions and organic molecules from the external environment. Most fungal species are filamentous and possess cell walls made of chitin, the same substance that is found in insect exoskeletons.

Bryophytes are small (rarely more than 4 cm in height), very simple land plants, and the least abundant in terms of number of species and overall population. Bryophytes do not appear to be in the direct line of evolution leading to the vascular plants; rather, they seem to constitute a separate minor branch.

Bryophytes include mosses, liverworts, and hornworts. These small plants have life cycles that depend on water during the sexual phase. Water facilitates fertilization, the fusion of gametes to produce a diploid zygote, a feature also seen in the algal precursors of these plants.

Bryophytes are like algae in other respects as well: They have neither true roots nor true leaves, they lack a vascular system, and they produce no hard tissues for structural support. The absence of these structures that are important for growth on land greatly restricts the potential size of bryophytes, which, unlike algae, are terrestrial rather than aquatic.

The ferns represent the largest group of spore-bearing vascular plants. In contrast to the bryophytes, ferns have true roots, leaves, and vascular tissues, and they produce hard tissues for support.

These architectural features enable ferns to grow to the size of small trees. Although ferns are better adapted to the drying conditions of terrestrial life than bryophytes are, they still depend on water as a medium for the movement of sperm to the egg. This dependence on water during a critical stage of their life cycle restricts the ecological range of ferns to relatively moist habitats.

The most successful terrestrial plants are the seed plants. Seed plants have been able to adapt to an extraordinary range of habitats. The embryo, protected and nourished inside the seed, is able to survive in a dormant state during unfavourable growing conditions such as drought. Seed dispersal also facilitates the dissemination of the embryos away from the parent plant.

Another important feature of seed plants is their mode of fertilization. Fertilization in seed plants is brought about by wind- or insect-mediated transfer of pollen, the gamete-producing structure of the male, to the sexual structure of the female, the pistil. Pollination is independent of external water, a distinct advantage in terrestrial environments. Many seed plants produce massive amounts of woody tissues, which enable them to grow to extraordinary heights. These features of seed plants have contributed to their success and account for their wide range.

There are two categories of seed plants: gymnosperms (from the Greek for "naked seed") and angiosperms (based on the Greek for "vessel seed," or seeds contained in a vessel). Gymnosperms are the less advanced type; about 700 species of gymnosperms are known. The largest group of gymnosperms is the conifers ("cone-bearers"), which include such commercially important forest trees as pine, fir, spruce, and redwood.

Two types of cones are present: male cones, which produce pollen, and female cones, which bear ovules. The ovules are located on the surfaces of specialized structures called cone scales. After wind-mediated pollination, the sperm reaches the egg via a pollen tube, and the fertilized egg develops into an embryo. Upon maturation, the cone scales, which are appressed during early development, separate from each other, allowing the naked seeds to fall to the ground.

Angiosperms, the more advanced type of seed plant, first became abundant during the Cretaceous period, about 100 million years ago. Today, they dominate the landscape, easily outcompeting their cousins, the gymnosperms. About 250,000 species are known, but many more remain to be characterized. A typical angiosperm life cycle, that of *Zea mays* (corn). The major innovation of the angiosperms is the flower; hence they are referred to as *flowering plants*. There are other anatomical differences between angiosperms and gymnosperms, but none so crucial and far-reaching as the mode of reproduction.

Flower Structure and the Angiosperm Life Cycle

The flower consists of several leaflike structures attached to a specialized region of the stem called the receptacle. Sepals and petals are the most leaflike. Petals have the primary function of attracting insects to serve as pollinators, accounting for their often showy and brightly coloured appearance.

The stamen is the male sexual structure, and the pistil is the female sexual structure. The pistil is composed of one or more united carpels; the pistil, or in some flowers a whorl of pistils, is sometimes referred to as the gynoecium. The stamen consists of a narrow stalk called the filament and a chambered structure called the anther. The anther contains tissue that gives rise to pollen grains. The pistil consists of the stigma (the tip where pollen lands during pollination), the style (an elongated structure), and the ovary. The ovary, the hollow basal portion of the pistil, completely encloses one or more ovules. Each ovule, in turn, contains an embryo sac, the structure that gives rise to the female gamete, the egg.

After landing on the stigma, the pollen grain germinates to form a long pollen tube, which penetrates the tissues of the style and ultimately enters the cavity of the ovary, which houses the ovule. Within the ovary, the pollen tube enters the ovule and deposits two haploid sperm cells in the embryo sac. One sperm cell fuses with the egg to produce the zygote; the other typically fuses with the two polar nuclei to produce a specialized storage tissue termed the endosperm, which provides nutrients to the growing embryo.

Endosperm tissue also provides the bulk of the worldrquotes food supply in the form of cereal grains. As in conifers, in angiosperms the outer tissues of the ovule harden into a protective seed coat. Angiosperm seeds have a second layer of protective tissues, the fruit. The fruit consists of the ovary wall and, in some cases, receptacle tissue.

Angiosperms are divided into two major groups, dicotyledons (dicots) and monocotyledons (monocots). This distinction is based primarily on the number of cotyledons, or seed leaves. In addition, the two groups differ with respect to other anatomical features, such as the arrangement of their vascular tissues, and their floral structure.

As the dominant plant group on Earth, and because of their great economic and agricultural importance, angiosperms have been studied much more intensively than other types of plants. Plant physiologists have focused on a relatively small number of species that represent convenient experimental systems for the study of specific phenomena.

Therefore, while we focus on these famous few, it is important to keep in mind the tremendous diversity of form and function that exists within the angiosperms, and the even greater diversity of form and function that is found within the plant kingdom as a whole.

Plant Tissue Systems: Dermal, Ground, and Vascular

Dermal tissue. The epidermis is the dermal tissue of young plants undergoing primary growth. It is generally composed of specialized, flattened polygonal cells that occur on all plant surfaces. Shoot surfaces are usually coated with a waxy cuticle to prevent water loss and are often covered with hairs, or trichomes, which are epidermal cell extensions.

Pairs of specialized epidermal cells, the guard cells, are found surrounding microscopic pores in all leaves. The guard cells and pores are called stomata (singular stoma), and they permit gas exchange (water loss, CO_2 uptake, and O_2 release or uptake) between the atmosphere and the interior of the leaf.

The root epidermis is adapted for absorption of water and minerals, and its outer wall surface typically does not have a waxy cuticle. Extensions from the root epidermal cells, the root hairs, increase the surface area over which absorption can take place.

Ground tissue. Making up the bulk of the plant are cells termed the ground tissue. There are three types of ground tissue: parenchyma, collenchyma, and sclerenchyma.

- Parenchyma, the most abundant ground tissue, consists of thin-walled, metabolically active cells that carry out a variety of functions in the plant including photosynthesis and storage
- Collenchyma tissue is composed of narrow, elongated cells with thick primary walls. Collenchyma cells provide structural support to the growing plant body, particularly shoots, and their thickened walls are nonlignified so they can stretch as the organ elongates. Collenchyma cells are typically arranged in bundles or layers near the periphery of stems or leaf petioles.
- Sclerenchyma consists of two types of cells, sclereids and fibres Both have thick secondary walls and are frequently dead at maturity. Sclereids occur in a variety of shapes, ranging from roughly spherical to branched, and are widely distributed throughout the plant. In contrast, fibres

are narrow, elongated cells that are commonly associated with vascular tissues. The main function of sclerenchyma is to provide mechanical support particularly to parts of the plant that are no longer elongating.

In the stem, the pith and the cortex make up the ground tissue. The pith is located within the cylinder of vascular tissue, where it often exhibits a spongy texture because of the presence of large intercellular air spaces.

If the growth of the pith fails to keep up with that of the surrounding tissues, the pith may degenerate, producing a hollow stem. In general, roots lack piths, although there are exceptions to this rule. In contrast, the cortex, which is located between the epidermis and the vascular cylinder, is present in both stems and roots.

At the boundary between the ground tissue and the vascular tissue in roots, and occasionally in stems, is a specialized layer of cortex known as the endodermis. This single layer of cells originates from cortical tissue at the innermost layer of the root cortex and forms a cylinder that surrounds the central vascular tissue, or stele.

Early in root development, a narrow band composed of the waxy substance suberin is formed in the cell walls circumscribing each endodermal cell. These suberin deposits, called Casparian strips, form a barrier in the endodermal walls to the intercellular movement of water, ions, and other water-soluble solutes to the vascular cells. Leaves have two interior layers of ground tissue that are collectively known as the mesophyll.

The palisade parenchyma consists of closely spaced, columnar cells located beneath the upper epidermis. There is usually one layer of palisade parenchyma in the leaf. Palisade parenchyma cells are rich in chloroplasts and are a primary site of photosynthesis in the leaf. Below the palisade parenchyma are i rregularly shaped, widely spaced spongy mesophyll cells.

The spongy mesophyll cells are also photosynthetic, and the large spaces between these cells allow diffusion of carbon dioxide. The spongy mesophyll also contributes to leaf flexibility in the wind, and this flexibility facilitates the movement of gases within the leaf.

Vascular tissues: xylem and phloem. The vascular tissue is composed of two major conducting systems: the xylem and the phloem. The xylem transports water and mineral ions from the root to the rest of the plant.

The phloem distributes the products of photosynthesis and a variety of other solutes throughout the plant. The tracheids and vessel elements are the conducting cells of the xylem.

Both of these cell types have elaborate secondary-wall thickenings and lose their cytoplasm at maturity; that is, they are dead when functional. Tracheids overlap each other, whereas vessel elements have open end walls and are arranged end to end to form a larger unit called a vessel. Other cell types present in the xylem include parenchyma cells, which are important for the storage of energy-rich molecules and phenolic compounds, and sclerenchyma fibres.

The sieve elements and sieve cells are responsible for sugar translocation in the phloem. The former are found in angiosperms; the latter perform the same function in gymnosperms. Like vessel elements, sieve elements are often stacked in vertical rows, forming larger units called sieve tubes, whereas sieve cells form overlapping arrays. Both types of conducting cells are living when functional, but they lack nuclei and central vacuoles and have relatively few cytoplasmic organelles.

Substances are translocated from sieve cell to sieve cell laterally through circular or oval zones containing enlarged pores, called sieve areas. In contrast, sieve tubes translocate substances through large pores in the end walls of the sieve elements, called sieve plates. Sugar movement through sieve tubes is more efficient and rapid than through sieve cells and represents a more evolutionarily advanced mechanism.

Sieve elements are associated with, and depend on, densely cytoplasmic parenchyma cells called companion cells. The analogous cells adjacent to the sieve cells of gymnosperms are called albuminous cells. Companion cells provide proteins and metabolites necessary for the functions of the sieve tube elements. In addition, the phloem frequently contains storage parenchyma and fibres that provide mechanical support.

The Structures of Chloroplast Glycosylglycerides

The structures of the three plant glycosylglycerides, monogalactosyldiacylglycerol (MGDG), digalactosyldiacylglycerol (DGDG) and sulfoquinovosyldiacylglycerol (SQDG). The two galactolipids, MGDG and DGDG, are uncharged but polar; the sulfolipid, SQDG, is negatively charged as well as being polar.

The Multiple Steps in Construction of the Cell Plate Following Mitosis

During Mitosis in plants, the daughter chromosomes separate during anaphase and migrate toward their respective poles. In contrast to animal cells, the spindle poles of plant cells is quite diffuse, so that the chromosomes move in nearly parallel fashion.

At the end of teleophase, cytokinesis begins with the formation of the cell plate. This process involves fusion of many small secretory vesicles, and attachment of the resulting structure to the plasma membrane. A diagram of the steps involved in cell plate formation.

In the first step, Golgi vesicles, some of which are interconnected via *fusion tubes,* aggregate in the spindle midzone area. This structure is called the fusion tube network (FTN). The transition from the first to the second stage of cell plate formation can be inhibited by caffeine. (B) Formation of a tubulo-vesicular network (TVN).

The contents of the vesicles, mainly pectins, represent the precursors from which the new middle lamella is assembled outside the cell. In the next stage, vesicle fusion increases, forming a tubulo-vesicular network (TVN), and the membranes become coated with either clathrin or other proteins. (C) In the third stage, the central region of the growing cell plate forms a tubular network (TN), with vesicle fusion occurring at the growing edges where the remaining microtubules are located. (D) In the final stage, the cell plate contacts and adheres to the plasma membrane of the parent cell.

At the same time the tubular network expands to form a fenestrated sheet. (E) At the end of mitosis, the phragmoplast disappears, the cell enters interphase, and microtubules reappear in the cytosol near the plasma membrane, where they play a role in the deposition of cellulose microfibrils during cell wall growth.

(F) Electron micrograph of a cell plate forming in a root tip of a beet, (*Beta vulgaris*) (10,000×) MT, microtubule; VE, vesicles; N, nucleus; NE, nuclear envelope; P, cell plate.

Water and Plant Cells

Calculating Capillary Rise

We learn on page 36 of the textbook that the water properties of cohesion, adhesion and surface tension, give rise to the phenomenon of capillarity, the movement of water for small distances up a capillary tube. The smaller the tube radius, the higher the capillary rise. How far the water will move may be calculated using the following formula:

$$\text{Capillary rise} = \frac{14.9 \times 10^{-6}\,\text{m}^2}{\text{radius}}$$

where both capillary rise and radius are expressed in meters.

For a xylem vessel with 25 μm radius, the capillary rise is about 0.6 m. This distance is much too small to be significant for water transport up tall trees.

Fibrous materials such as cell walls can act like wicks to draw water by capillarity from nearby xylem.

This capillary action ensures that cell wall surfaces that are directly exposed to the air, such as those in leaf mesophyll, remain wetted and do not dry out. Because the cell wall capillaries have a tiny radius, about 10 m^{-8}, very large physical forces can be generated in the water just below the evaporative surfaces of cell walls.

Calculating Half-Times of Diffusion

Extremely Slow over Long Distances

From Fick's law, one can derive an expression for the time it takes for a substance to diffuse a particular distance.

If one defines conditions such that all the solute molecules are concentrated at the starting position, then the concentration front moves away from the starting position, as shown for a later time point. As the substance diffuses away from the starting point, the concentration gradient becomes less steep (Dc_s decreases) and thus net movement becomes slower.

The time it takes for the substance at any given distance from the starting point to reach one-half of the concentration at the starting point (t_c = ½) is given by the following equation:

$$t_{c=1/2} = \frac{(\text{dis tan ce})^2}{D_S} K$$

where K is a constant and D_s is the diffusion coefficient. The above equation shows that the time required for a substance to diffuse a given distance increases in proportion to the *square* of that distance. Let us consider two numerical examples. First, how long it would take a small molecule to diffuse across a typical cell? The diffusion coefficient for a small molecule like glucose is about 10^{-9} m^2 s^{-1}, and the cell length may be 50 µm. Thus, for this example:

$$t_{c=1/2} = \frac{(50+10^{-6}\,\text{m})^2}{10^{-9}\,\text{m}^2\text{s}^{-1}} = 2.5\text{s}$$

This calculation shows that small molecules diffuse over cellular dimensions rapidly. What about diffusion over longer distances? Calculating the time needed for the same substance to move a distance of 1 m (e.g., the length of a corn leaf), we find:

$$t_{c=1/2} = \frac{(1\text{m})^2}{10^{-9}\,\text{m}^2\text{s}^{-1}} = 10^9\,\text{s} \approx 32\text{ year}$$

a value that exceeds by orders of magnitude the life span of a corn plant, which lives only a few months. This shows that diffusion in solutions can be effective within cellular dimensions but is far too slow for mass transport over long distances. Diffusion is of great importance as a driving force for the water vapour lost from leaves, because the diffusion coefficient for a molecule in air is much greater than in aqueous solutions.

Alternative Conventions for Components of Water Potential

Students planning further study of plant water relations should note that the components of water potential defined in the text are sometimes given different names and symbols. In particular, the equation

$$\Psi_w = \Psi_s + \Psi_p$$

is often replaced by the following equivalent equation:

$$\psi_w = \pi + P$$

In this alternative convention, *P* is the same as Y_p. It is the hydrostatic pressure of the solution, and may be positive, as in turgid cells, or negative, as in xylem water.

The symbol p is called osmotic pressure and is the negative of Y_s. That is, p has positive values, and Y_s has negative values. "Osmotic pressure" is the term that physical chemists, zoologists, and many others use to denote the effect of dissolved solutes on the free energy of water. Most handbooks of physics and chemistry use the term "osmotic pressure" and the symbol π.

The negative sign in front of p in the equation above accounts for the reduction in water potential (Y_w) by dissolved solutes. Thus $Y_s = -$ p. A very interesting, if somewhat unconventional, account of the history and physical meaning of osmotic pressure is given by Hammel and Scholander (1976).

Unfortunately, some authors have mixed the conventions for Y_s and p, leading to unnecessary confusion about what is meant by the symbol p. Thus, p is sometimes incorrectly called osmotic potential instead of osmotic pressure, and it may be used either as a positive quantity or as a negative quantity.

Regarding matric potential, it is usually designated by the symbol *t*.

Can Negative Turgor Pressures Exist in Living Cells?

One of the challenging aspects of understanding plant water relations is the remarkable range of pressures, both positive and negative, that occur within the bodies of plants.

Negative pressures (or tensions), which depend upon the cohesive strength of water coupled with the strength of lignified cell walls to resist deformation, play an important role in water transport through the xylem. Positive pressures, which depend upon the semipermeable nature of the plasma membrane and the elastic nature of primary cell walls, occur in all hydrated living plants cells but can be especially large in sieve tubes and guard cells.

Living plant cells are typically assumed to have only positive pressures. However, there appears to be no reason that negative pressures could not also occur within the cytoplasm of living plant cells. This web topic explores the existence and potential role of negative turgor pressures in plants.

When a cell loses water in air, turgor declines and solute concentrations increase. At the turgor loss point (y_p = 0), the hydrostatic pressure in the cell sap is equal to atmospheric pressure, meaning that no net force is exerted on the cell wall.

If water continues to be lost from the cell, the pressure within the cytoplasm drops below atmospheric pressure, resulting in a force imbalance that collapses the cell wall. The deformation of living cells upon desiccation is called cytorrhysis. Note that the plasma membrane is pressed against the cell wall throughout desiccation (i.e., plasmolysis does not occur) because the hydrostatic pressure in cytoplasm remains greater than the hydrostatic pressure in apoplast.

In the living cells described above, a decrease in cell water potential below the turgor loss point is balanced by an increase in solute concentration as the volume of the cell decreases.

Thus, true "negative" pressures do not develop. In contrast, xylem conduits have rigid cell walls that resist deformation, allowing them to sustain negative pressures without imploding. This raises the question of what happens when water is lost from living cells that have thick walls or are embedded in a rigid matrix of cells (e.g., xylem parenchyma). Might these cells resist deformation and thus develop negative turgor pressures?

It is important to emphasize that this is a controversial area, due in large part to the absence of direct measurements of negative turgor pressures within cells. However, before reviewing the evidence for negative turgor pressures, it is worthwhile to consider what physiological effects might result from the development of such negative pressures in living cells.

The major outcome of negative turgor is that it allows stiff-walled cells to decrease in water potential without undergoing major changes in cell volume or osmotic concentration. Because cytorrhysis might cause physical damage to the wall and/or cell membranes, while the high concentrations of solutes resulting from the reduction in cell volume might adversely affect the conformation of membranes and proteins, it is possible to imagine a physiological role or benefit for the development of negative turgor pressures in plant cells.

We now turn this around and ask if there are any downsides to the generation of negative turgor pressures in living cells? In the xylem, the primary issue associated with negative pressures is cavitation; could this also be a concern for living cells?

The primary mechanism for cavitation in plants is air seeding, reflecting the fact that the probability of the de novo formation of gas voids in water (either by homogeneous or heterogeneous nucleation) is extremely low. In air seeding, air is sucked in through the cell wall, where it then "nucleates" the transition to the vapour phase.

For air seeding to occur in a living cell experiencing negative pressures, air would have to be pulled through the very small pores of the cell wall. While one can imagine this happening, the movement of air across the cell wall would result in plasmolysis and thus the immediate release of any tension in the cytoplasm because the plasma membrane is not capable of withstanding any significant pressure.

Other costs associated with the ability of living cells to generate negative pressures include the metabolic costs of producing rigid cell walls, as well as any limitations lignified walls might place on physiological function. In addition, a strategy for avoiding cell damage due to desiccation via the generation of negative pressures might impose limitations on cell size. The strength of a cell to withstand collapse (and thus generate negative pressures) is inversely proportional to cell size.

Evidence for negative turgor pressure in plants is limited, in part reflecting the fact that few researchers have devoted much attention to this issue. Living cells with flexible (unlignified) cell walls deform relatively easily upon desiccation and thus do not appear to support negative pressures. However, measurements of the forces needed to collapse cell walls suggest that living cells with thick walls can withstand forces in the range of 1.0 MPa.

Visual examination of tissues adapted to withstand very cold temperatures also provides indirect evidence for negative turgor. For example, frozen ray parenchyma cells often do not exhibit significant deformation, despite the very strong desiccatory effects (low water potential) of extracellular ice. One can hypothesize that the existence of negative pressures within these cells acts to balance the water potential gradient across the cell membrane (i.e., between the cytoplasm and ice formed within the apoplast).

In summary, negative turgor pressure remains an intriguing but little-studied area of plant water relations.

While it does not appear to form in cells such as the leaf mesophyll, that can deform easily, its existence in living cells with either lignified cell walls or where the cell is embedded in a matrix of lignified tissues (e.g., living cells within wood) cannot be ruled out. The potential benefits of negative turgor pressures in terms of preventing mechanical and osmotic damage associated with severe desiccation makes.

The Matric Potential

In discussions of soils, seeds, and cell walls, one often finds reference to yet another component of Y_w: the matric potential (Y_m).

The matric potential is used to account for the reduction in free energy of water when it exists as a thin surface layer, one or two molecules thick, adsorbed onto the surface of relatively dry soil particles, cell walls, and other materials.

The matric potential does not represent a new force acting on water, because the effect of surface interactions can theoretically be accounted for by an effect on Y_s and Y_p. In dry materials, however, this surface interaction effect often cannot easily be separated into Y_p and Y_s components in dry materials, so they are frequently bulked together and designated as the matric potential.

It is generally not valid to add Y_m to independent measurements of Y_s and Y_p to arrive at a total water potential.

This is particularly true for water inside hydrated cells and cell walls, where matric effects are either negligible or they are accounted for by a reduction in Y_p.

For instance, the negative pressure in water held by cell wall microcapillaries at the evaporative surfaces of leaves, is sometimes described as a wall matric potential. Care is needed to avoid inconsistencies when accounting for this physical effect in definitions of Y_p, Y_s, and Y_m.

Measuring Water Potential

Plant scientists have expended considerable effort in devising accurate and reliable methods for evaluating the water status of a plant. Four instruments that have been used extensively to measure Ψ_w, Ψ_s, and Ψ_p are described here: psychrometer, pressure chamber, cryoscopic osmometer, and pressure probe.

Psychrometer (Y_w Measurement)

Psychrometry (the prefix "psychro-" comes from the Greek word *psychein*, "to cool") is based on the fact that the vapour pressure of water is lowered as its water potential is reduced. Psychrometers measure the water vapour pressure of a solution or plant sample, on the basis of the principle that evaporation of water from a surface cools the surface.

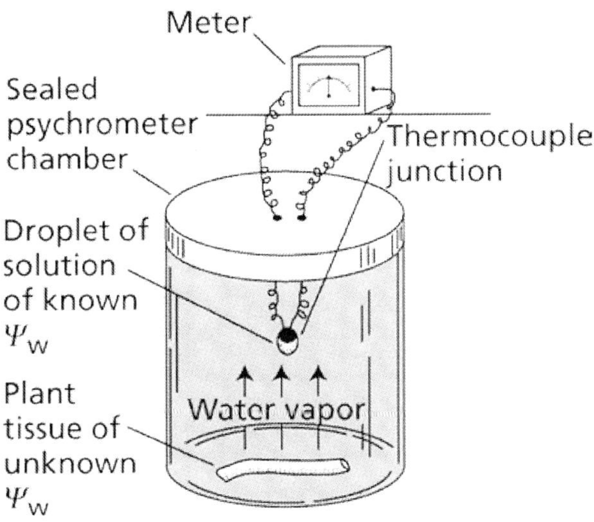

Figure: *Diagram illustrating the use of isopiestic psychrometry to measure the water potential of a plant tissue.*

One psychrometric technique, known as *isopiestic psychrometry*, has been used extensively by John Boyer and coworkers and is illustrated in above Figure. Investigators make a measurement by placing a piece of tissue sealed inside a small chamber that contains a temperature sensor (in this case, a thermocouple) in contact with a small droplet of a standard solution of known solute concentration (known Y_s and thus known Y_w).

If the tissue has a lower water potential than that of the droplet, water evaporates from the droplet, diffuses through the air, and is absorbed by the tissue. This slight evaporation of water cools the drop.

The larger the difference in water potential between the tissue and the droplet, the higher the rate of water transfer and hence the cooler the droplet. If the standard solution has a lower water potential than that of the sample to be measured, water will diffuse from the tissue to the droplet, causing warming of the droplet.

Measuring the change in temperature of the droplet for several solutions of known Y_w makes it possible to calculate the water potential of a solution for which the net movement of water between the droplet and the tissue would be zero signifying that the droplet and the tissue have the same water potential.

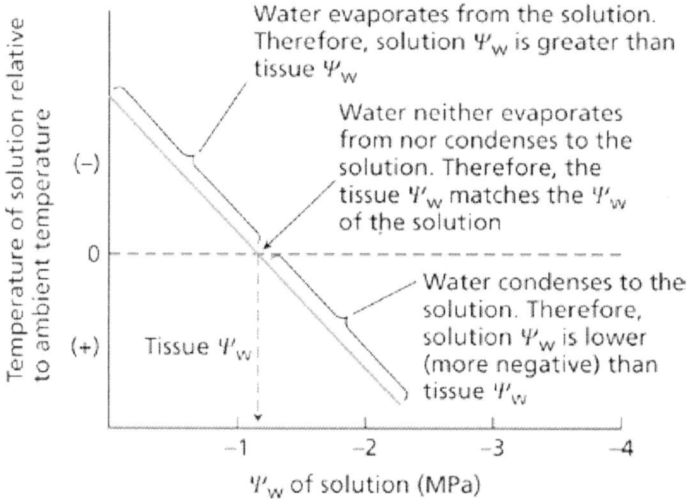

Figure: *The temperature of the sensor in the psychrometer shown in above figure depends on the water potential of the solution relative to the water potential of the tissue sample.*

Psychrometers can be used to measure the water potentials of both excised and intact plant tissue. Moreover, the method can be used to measure the Y_s of solutions.

This can be particularly useful with plant tissues. For example, the Y_w of a tissue is measured with a psychrometer, and then the tissue is crushed and the Y_s value of the expressed cell sap is measured with the same instrument. By combining the two measurements, researchers can estimate the turgor pressure that existed in the cells before the tissue was crushed ($Y_p = Y_w - Y_s$).

A major difficulty with this approach is the extreme sensitivity of the measurement to temperature fluctuations. For example, a change in temperature of 0.01°C corresponds to a change in water potential of about 0.1 MPa. Thus, psychrometers must be operated under constant temperature conditions. For this reason, the method is used primarily in laboratory settings. There are many variations in psychrometric technique; interested readers should consult Brown and Van Haveren 1972, Slavik 1974, and Boyer 1995.

Pressure Chamber

A relatively quick method for estimating the water potential of large pieces of tissues, such as leaves and small shoots, is by use of the pressure chamber. This method was pioneered by Henry Dixon at Trinity College, Dublin, at the beginning of the twentieth century, but it did not come into widespread use until P. Scholander and coworkers at the Scripps Institution of Oceanography improved the instrument design and showed its practical use.

In this technique, the organ to be measured is excised from the plant and is partly sealed in a pressure chamber.

Before excision, the water column in the xylem is under tension. When the water column is broken by excision of the organ (i.e., its tension is relieved allowing its Y_p to rise to zero), water is pulled rapidly from the xylem into the surrounding living cells by osmosis.

The cut surface consequently appears dull and dry. To make a measurement, the investigator pressurizes the chamber with compressed gas until the distribution of water between the living cells and the xylem conduits is returned to its initial, pre-excision, state.

This can be detected visually by observing when the water returns to the open ends of the xylem conduits that can be seen in the cut surface. The pressure needed to bring the water back to its initial distribution is called the *balance pressure* and is readily detected by the change in the appearance of the cut surface, which becomes wet and shiny when this pressure is attained.

The pressure chamber is often described as a tool to measure the tension in the xylem.

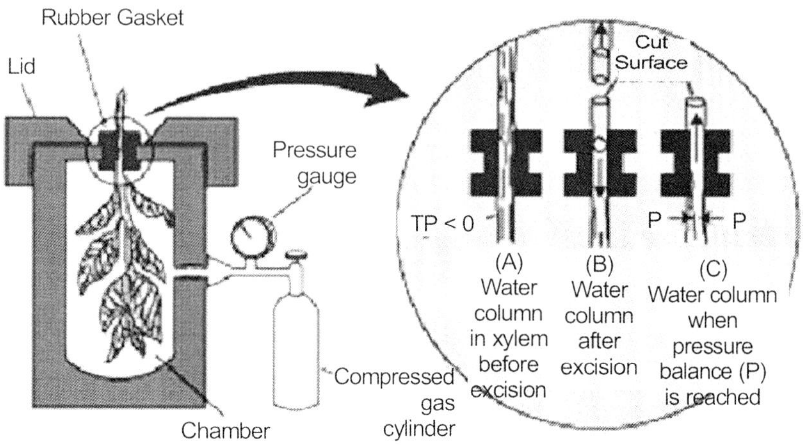

Figure: *The pressure chamber method for measuring plant water potential. The diagram at left shows a shoot sealed into a chamber, which may be pressurized with compressed gas. The diagrams at right show the state of the water columns within the xylem at three points in time: (A) The xylem is uncut and under a negative pressure, or tension. (B) The shoot is cut, causing the water to pull back into the tissue, away from the cut surface, in response to the tension in the xylem. (C) The chamber is pressurized, bringing the xylem sap back to the cut surface.*

However, this is only strictly true for measurements made on a non-transpiring leaf or shoot (for example, one that has been previously enclosed in a plastic bag). When there is no transpiration, the water potential of the leaf cells and the water potential in the xylem will come into equilibrium.

The balancing pressure measured on such a non-transpiring shoot is equal in magnitude but opposite in sign to the pressure in the xylem (Y_p). Because the water potential of our non-transpiring leaf is equal to the water potential of the xylem, one can calculate the water potential of the leaf by adding together Y_p and Y_s of the xylem, provided one collects a sample of xylem sap for determination of Y_s.

Luckily Y_s of the xylem is usually small (> -0.1 MPa) compared to typical midday tensions in the xylem (Y_p of –1 to –2 MPa). Thus, correction for the Y_s of the xylem sap is frequently omitted.

Balancing pressure measurements of transpiring leaves are more difficult to interpret. The fact that water is flowing from the xylem to the leaf means that differences in water potential must exist.

When the transpiring leaf or shoot is cut off, the tension in the xylem is instantly relieved and water is drawn into the leaf cells until the water potentials of the xylem and the leaf cells come into equilibrium. Because the total volume of the leaf cells is much larger than the volume of sap in the xylem, this equilibrium water potential will be heavily weighted towards that of the leaf.

Thus, any measurement of the balancing pressure on such a leaf or shoot will result in a value that is approximately the water potential of the leaf, rather than the tension of the xylem. (To be exact, one would have to add the Y_s of the xylem sap to the negative of the balancing pressure to get the leaf water potential.) One can explore the differences between the water potential of the xylem and the water potential of a transpiring leaf by comparing balancing pressures measured on covered (i.e., non-transpiring) versus uncovered (transpiring) leaves.

Pressure chamber measurements provide a quick and accurate way of measuring leaf water potential. Because the pressure chamber method does not require delicate instrumentation or temperature control, it has been used extensively under field conditions. For a more complete description of the theory and operation of the pressure chamber.

Cryoscopic Osmometer

The cryoscopic osmometer measures the osmotic potential of a solution by measuring its freezing point.

Solutions have *colligative* properties that collectively depend on the number of dissolved particles and not on the nature of the solute. For example, solutes reduce the vapour pressure of a solution, raise its boiling point, and lower its freezing point.

The specific nature of the solute does not matter. One of the colligative *properties* of solutions is the decrease in the freezing point as the solute concentration increases. For example, a solution containing 1 mol of solutes per kilogram of water has a freezing point of –1.86°C, compared with 0°C for pure water.

Various instruments can be used to measure the freezing-point depression of solutions. With a cryoscopic osmometer, solution samples as small as 1 nanoliter (10^{-9} L) are placed in an oil medium located on the temperature-controlled stage of a microscope. The very small sample size allows sap from single cells to be measured and permits rapid thermal equilibration with the stage.

To prevent evaporation, the investigator suspends the samples in oil-filled wells in a silver plate (silver has high thermal conductivity).

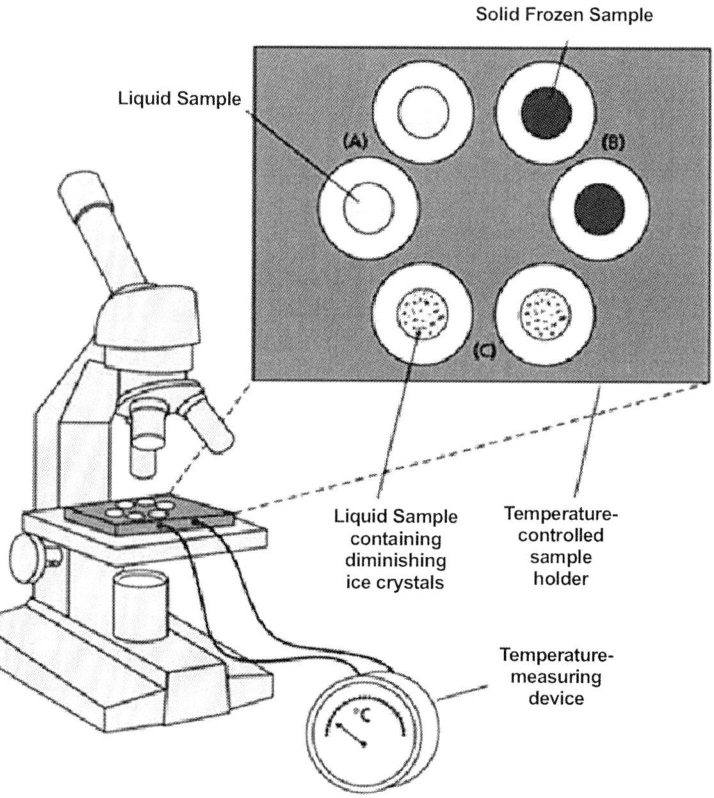

***Figure:** A cryoscopic osmometer measures the concentration of total dissolved solutes by measuring the freezing-point depression of a solution. (A) Very small liquid samples are loaded onto the temperature-controlled stage of a microscope. (B) When the temperature is quickly reduced, the samples supercool and freeze. (C) Slowly warming the stage causes the samples to thaw. The temperature at which the last ice crystal melts provides a measure of the melting point of the sample.*

The temperature of the stage is rapidly decreased to about –30° C, which causes the sample to freeze. The temperature is then raised very slowly, and the melting process in the sample is observed through the microscope. When the last ice crystal in the sample melts, the temperature of the stage is recorded (note that the melting and freezing points are the same).

It is straightforward to calculate the solute concentration from the freezing-point depression; and from the solute concentration (c_s), Y_s is calculated as $-RTc_s$. This technique has been used to measure droplets extracted from single cells.

Pressure Probe

If a cell were as large as a watermelon or even a grape, measuring its hydrostatic pressure would be a relatively easy task. Because of the small size of plant cells, however, the development of methods for direct measurement of turgor pressure has been slow.

Using a micromanometer, Paul Green at the University of Pennsylvania developed one of the first direct methods for measuring turgor pressure in plant cells (Green and Stanton 1967). In this technique, an air-filled glass tube sealed at one end is inserted into a cell.

The high pressure in the cell compresses the trapped gas, and from the change in volume one can readily calculate the pressure of the cell from the ideal gas law (pressure × volume = constant). This method works only for cells of relatively large volume, such as the giant cell of the filamentous green alga *Nitella*. For smaller cells, the loss of cell sap into the glass tube is sufficient to deflate the cell and this yields artifactually low pressures.

For higher plant cells, which are several orders of magnitude smaller in volume than *Nitella,* a more sophisticated device, the pressure probe, was developed by Ernest Steudle, Ulrich Zimmermann, and their colleagues in Germany. This instrument is similar to a miniature syringe.

A glass microcapillary tube is pulled to a fine point and is inserted into a cell. The microcapillary is filled with silicone oil, a relatively incompressible fluid that can be readily distinguished from cell sap under a microscope.

When the tip of the microcapillary is first inserted into the cell, cell sap begins to flow into the capillary because of the initial low pressure of that region. Investigators can observe such movement of sap under the microscope and counteract it by pushing on the plunger of the device, thus building up a pressure. In such fashion the boundary between the oil and the cell sap can be pushed back to the tip of the microcapillary.

When the boundary is returned to the tip and is held in a constant position, the initial volume of the cell is restored and the pressure inside the cell is exactly balanced by the pressure in the capillary. This pressure is measured by a pressure sensor in the device. Thus the hydrostatic pressure of individual cells may be measured directly.

This method has been used to measure Y_p and other parameters of water relations in cells of both excised and intact tissues of a variety of plant species. The primary limitation of this method is that some cells are too small to measure.

Furthermore, some cells tend to leak after being stabbed with the capillary, and others plug up the tip of the capillary, thereby preventing valid measurements. The pressure probe has also been adapted to measure positive and negative values of Y_p in the xylem. However, technical problems with cavitation limit the measurement of negative Y_p by this technique.

Understanding Hydraulic Conductivity

Consider a cell with an initial water potential of –0.2 MPa, submerged in pure water. From this information we know that water will flow into the cell and that the driving force is DY_w = 0.2 MPa, but what is the initial rate of movement? The rate depends on permeability of the membrane to water, a property usually called the hydraulic conductivity (*Lp*) of the membrane.

Driving force, membrane permeability, and flow rate are related by the following equation:

Flow rate = driving force × hydraulic conductivity

Hydraulic conductivity expresses how readily water can move across a membrane and has units of volume of water per unit area of membrane per unit time per unit driving force (for instance, $m^3\ m^{-2}\ s^{-1}\ MPa^{-1}$ or $m\ s^{-1}\ MPa^{-1}$). The larger the hydraulic conductivity, the larger the flow rate. The hydraulic conductivity of the membrane is 10^{-6} m s^{-1} MPa^{-1}. The transport (flow) rate (J_v) can then be calculated from the following equation:

$$J_v = (Lp(\Delta\psi_w)$$

where J_v is the volume of water crossing the membrane per unit area of membrane and per unit time (m^3 m^{-2} s^{-1} or, equivalently, m s^{-1}). Please note that this equation assumes that the membrane is ideal—that is, that solute transport is negligible and water transport is equally sensitive to DY_s and DY_p across the membrane. Nonideal membranes require a more complicated equation that separately accounts for water flow induced by DY_s and by DY_p.

In our example, J_v has a value of 0.2×10^{-6} m s^{-1}. Note that J_v has the physical meaning of a velocity. We can calculate the flow rate in volumetric terms (m^3 s^{-1}) by multiplying J_v by the surface area of the cell.

The resulting value is the *initial* rate of water transport. As water is taken up, cell Y_w increases and the driving force (DY_w) decreases. As a result, water transport slows with time. As elaborated in p. 43 of the textbook, the rate approaches zero in an exponential manner.

Wilting and Plasmolysis

Plasmolysis is the separation of plant cell cytoplasm from the cell wall as a result of water loss. It is unlikely to occur in nature, except in severe conditions. Plasmolysis is induced in the laboratory by immersing a plant cell in a strongly saline or sugary solution, so that water is lost by osmosis.

If onion epidermal tissue is immersed in a solution of calcium nitrate, cells rapidly lose water by osmosis and the protoplasm of the cells shrinks. This occurs because the calcium and nitrate ions freely permeate the cell wall and encounter the selectively permeable plasma membrane.

The large vacuole in the centre of the cell originally contains a dilute solution with much lower osmotic pressure than that of the calcium nitrate solution on the other side of the membrane. The vacuole thus loses water and becomes smaller. The space between the cell membrane and the cell wall enlarges and the plasma membrane and the protoplasm within it contract to the centre of the cell. Strands of cytoplasm extend to the cell wall because of plasma membrane-cell wall attachment points. Plasmolysed cells die unless they are transferred quickly from the salt or sugar solution to water.

Mineral Nutrition

Symptoms of Deficiency in Essential Minerals

Visual nutrient deficiency symptoms can be a very powerful diagnostic tool for evaluating the nutrient status of plants. One should keep in mind, however, that a given individual visual symptom is seldom sufficient to make a definitive diagnosis of a plant's nutrient status.

Many of the classic deficiency symptoms such as tip burn, chlorosis and necrosis are characteristically associated with more than one mineral deficiency and also with other stresses that by themselves are not diagnostic for any specific nutrient stress. However, their detection is extremely useful in making an evaluation of nutrient status.

In addition to the actual observations of morphological and spectral symptoms, knowing the location and timing of these symptoms is a critical aspect of any nutrient status evaluation. Plants do not grow in isolation, they are part of the overall environment and as such they respond to environmental changes as that affect nutrient availability. Also, plants do influence their environment and can contribute to environmental changes, which in turn can affect the nutrient status of the plant.

Sources of Visual Symptoms

Stresses such as salinity, pathogens, and air pollution induce their own characteristic set of visual symptoms. Often, these symptoms closely resemble those of nutrient deficiency.

Pathogens often produce an interveinal chlorosis, and air pollution and salinity stress can cause tip burn. Although at first these symptoms might seem similar in their general appearance to nutrient deficiency symptoms, they do differ in detail and/or in their overall developmental pattern.

Pathological symptoms can often be separated from nutritional symptoms by their distribution in a population of affected plants.

If the plants are under a nutrient stress, all plants of a given type and age in the same environment tend to develop similar symptoms at the same time.

However if the stress is the result of pathology, the development of symptoms will have a tendency to vary between plants until a relatively advanced stage of the pathology is reached.

Environmental Associations

Plants remove substantial amounts of nutrients from the soil during their normal growth cycle and many long-term environmental changes occur as a result of this process. Effects on the soil go considerably beyond the straight removal or depletion of nutrients. Charge balance must be maintained in the plant-soil system during nutrient uptake. Charge balance is usually achieved by the excretion of proton and/or hydroxyl ions by the plant to replace the absorbed nutrient cations or anions.

For example when plants are fertilized with ammonia, they acquire most of their nitrogen in the form of the ammonium cation, rather than from the usual nitrate anion. Because nitrate is the only anion used by the plant in large amounts, the net result of this change is that during normal nutrient uptake the proton excretion will far exceed that of hydroxyl ions.

In the case of vigorously growing plants, the amount of excreted protons can be sufficiently large as to decrease the pH of the soil by several pH units. Changes in soil pH of such magnitude can have large implications for a number of soil processes such as soil structure, nutrient availability and leaching of nutrients.

The immediate effect on the soil may be favourable for some plants, especially acid-loving plants, in that it tends to make iron more available. However, in the long run, lowering the soil pH can be deleterious to plants in that the availability of nutrients will change. A lower soil pH will allow micronutrients to be more readily leached from the soil profile, eventually resulting in deficiencies of nutrients such as Cu and Zn. Additionally, when the pH of the soil drops much below pH 5, the solubility of Al and Mn can increase to such an extent as to become toxic to most plant growth.

Plants are often thought of as passive in relation to the environment. However this is not always a valid assumption; for there are many plants that clearly manipulate their environment in a fashion that tends to makes certain nutrients more readily available.

For example, iron is a limiting nutrient in many agricultural areas, but it comprises about 3% of the average soil which, if available, would be far in excess of the needs of the average plant. Some plants actively excrete protons, and the resulting decrease in pH increases the solubility of iron in their environment. In addition, other plants excrete phytosiderophores that chelate the soil iron rendering it a more available form for the plants.

Pathways of Symptom Development

At first glance, it would appear that the distinction of deficiency symptoms for the 13 known essential mineral nutrients should be relatively simple. But such an assumption is incorrect. In fact, the deficiency symptoms are quite complex because each nutrient has a number of different biological functions and each function may have an independent set of interactions with a wide range of environmental parameters.

In addition, the expression of these symptoms varies for acute or chronic deficiency conditions. Acute deficiency occurs when a nutrient is suddenly no longer available to a rapidly growing plant. Chronic deficiency occurs when there is a limited but continuous supply of a nutrient, at a rate that is insufficient to meet the growth demands of the plant. Most of the classic deficiency symptoms described in textbooks are characteristic of acute deficiencies. The most common symptoms of low-grade, chronic deficiencies are a tendency towards darker green leaves and stunted or slow growth.

Typically most published descriptions of deficiency symptoms arise from experiments conducted in greenhouses or growth chambers where the plants are grown in hydroponics or in media where the nutrients are fully available. In these conditions, nutrients are readily available while present, but when a nutrient is depleted, the plant suddenly faces an acute deficiency. Thus, hydroponic studies favour the development of acute deficiencies.

In experiments designed to study micronutrient deficiency symptoms, micronutrients are usually omitted from the nutrient solution. Micronutrients are often present in the seed or as contaminants in the environment, so a plant of adequate size will exhaust these trace amounts of micronutrient and develop characteristic acute deficiency systems.

When deficiency symptoms of macronutrients are sought, the macronutrient is removed suddenly from a suitable sized rapidly growing plant. Alternatively the plant can initially be given a one-time supply of the nutrient that is sufficient for a limited amount of growth. Because macronutrients are continuously required in relatively large amounts by rapidly growing plants, the available nutrients will be rapidly depleted, resulting in an acute deficiency.

In natural systems, the plant encounters many degrees and types of stresses that result in different types of symptoms occurring over time. Perhaps the most common nutrient deficiency in natural environments is the case of a limited nutrient supply that is continuously renewed at a low rate from soil weathering processes. In such cases, the limited nutrient availability results in chronic nutrient deficiency symptoms.

Effect of Nutrient Mobility on Symptom Development

The interaction between nutrient mobility in the plant, and plant growth rate can be a major factor influencing the type and location of deficiency symptoms that develop.

For very mobile nutrients such as nitrogen and potassium, deficiency symptoms develop predominantly in the older and mature leaves. This is a result of these nutrients being preferentially mobilized during times of nutrient stress from the older leaves to the newer leaves near the growing regions of the plant. Additionally, mobile nutrients newly acquired by the roots are also preferentially translocated to new leaves and the growing regions.

Thus old and mature leaves are depleted of mobile nutrients during times of stress while the new leaves are maintained at a more favourable nutrient status.

The typical localization of deficiency symptoms of very weakly mobile nutrients such as calcium, boron, and iron is the opposite to that of the mobile nutrients; these deficiency symptoms are first displayed in the growing regions and new leaves while the old leaves remain in a favourable nutrient status. (This assumes that these plants started with sufficient nutrient, but ran out of nutrient as they developed).

In plants growing very slowly for reasons other than nutrition (such as low light) a normally limiting supply of a nutrient could, under these conditions, be sufficient for the plant to slowly develop, maybe even without symptoms. This type of development is likely to occur in the case of weakly mobile nutrients because excess nutrients in the older leaves will eventually be mobilized to supply newly developing tissues.

In contrast, a plant with a similar supply that is growing rapidly will develop severe deficiencies in the actively growing tissue such as leaf edges and the growing region of the plant. A classic example of this is calcium deficiency in vegetables such as lettuce where symptoms develop on the leaf margins (tip burn) and the growing region near the meristems.

The maximal growth rate of lettuce is often limited by the internal translocation rate of calcium to the growing tissue rather than from a limited nutrient supply in the soil.

When moderately mobile nutrients such as sulfur and magnesium are the limiting nutrients of the system, deficiency symptoms are normally seen over the entire plant. However the growth rate and rate of nutrient availability can make a considerable difference on the locations at which the symptoms develop. If the nutrient supply is marginal compared to the growth rate, symptoms will appear on the older tissue, but if the nutrient supply is very low compared to the growth rate, or the nutrient is totally depleted, the younger tissue will become deficient first.

Plant Competition and Induced Deficiencies

When the observed symptoms are the direct result of a nutrient deficiency, the actions needed for correction are relatively straightforward. However symptoms are often the result of interactions with other environmental factors limiting the availability of the nutrient whose symptoms are expressed. The classic instance is that of iron deficiency induced by an excess of heavy metals in the environment. Transition metals such as Cu, Zn Cr and Ni compete with Fe and each other for plant uptake. Competition for uptake is not specific to Fe and heavy metals but is true for all mineral nutrients that are chemically similar and have similar uptake mechanisms. For example if the availability of Cu or Zn is relatively less than that of Fe, then excessive concentrations of some other metal such as Ni or Cr will induce a deficiency of one of these nutrients rather than Fe. In the case of the macronutrients, excessive amounts of Mg will compete with K for uptake and can possibly induce a K deficiency. The barrenness of serpentine soils is the result of such competition, with the high Mg of these soils inducing a Ca deficiency. The toxicity of a low pH soil is another example of a basic nutrient deficiency. Low pH has a two-fold effect on soil nutrients: It enhances the leaching of cations, reducing their availability in the soil, and the relatively abundant protons in the soil compete with Ca and other cations for uptake. Thus, nutrient deficiencies can be induced by a number of different mechanisms often working in concert to limit the availability of a nutrient.

Nutrient Demand and Use Efficiency

Although all plants of the same species respond similarly to nutrient stress, plants of similar species will often show significant differences in their nutrient use efficiency. This results from differences in growth rate, root distribution, phase of development, and efficiency of nutrient uptake and utilization. This implies that in any given location, plants from one species may become nutrient-deficient, while those from another species growing in the same environment right next to them, may not show any deficiency symptoms.

Growth rate also affects nutrient status. When the nutrient supply is barely inadequate for growth under existing environmental conditions, many plants adjust their growth rate to match that supported by the available nutrient supply without displaying typical visual deficiency symptoms. Agricultural systems differ from natural systems in that crop plants have been selected primarily for rapid growth under low stress conditions.

This rapid growth rate results in a high nutrient demand by these plants and a higher incidence of nutrient deficiency unless supplemental fertilizers are supplied. It is not uncommon to find agricultural crops showing severe signs of nutrient stress, with native plants growing in the same area showing little or no indication of nutrient stress. In agriculture systems chronic deficiency symptoms develop mostly in crops with little or limited fertilization. Acute nutrient deficiency symptoms most often occur when new crops with a higher nutrient demand are introduced, or less productive lands are brought under cultivation for the production of rapidly growing crop plants.

Uniformity of Nutrient Status

Not all tissues of a plant are at the same nutrient status during times of stress. Leaves on the same plant that are exposed to different environmental conditions, (such as light), or those of different ages may have considerable differences in nutrient status. Mineral nutrients are for the most part acquired by the roots and translocated throughout the plant. The distance of any part of the plant to the roots will influence nutrient availability, particularly in the case of the less mobile nutrients. In plants recovering from nutrient deficiency, the root and conductive tissues recover first. For example, in the case of recovery from Fe deficiency, it is common to see the veins re-green while the interveinal tissue remains chlorotic and Fe-deficient.

In order to maintain rapid, optimal growth, all plant tissues must have a favourable nutrient status. Although a plant may be marginally low in a number of nutrients, only one nutrient at a time will limit overall growth. However, if the supply of that limiting nutrient is increased even slightly, the resulting increase in growth will increase the demand for all other nutrients and another nutrient, the next lowest in availability, will become limiting.

Other Diagnostic Tools

Although visual diagnostic symptoms are an extremely valuable tool for the rapid evaluation of the nutrient status of a plant, they are only some of the tools available. Other major tools include microscopic studies, spectral analysis, and tissue and soil analysis. These methods all vary in their precision, rapidity and their ability to predict future nutrient status. Because of the close interaction between plant growth and the environment, all predictions of future nutrient status must make assumptions about how the environment will change in that time frame.

The principle advantage of visual diagnostic symptoms is that they are readily obtained and provide an immediate evaluation of nutrient status. Their main drawback is that the visual symptoms do not develop until after there has been a major effect on yield, growth and development.

Tissue analysis is nutrient-specific but relatively slow; tissues must be sampled, processed and analysed before the nutrient status can be determined. An analysis of the mineral nutrient content of selected plants tissues, when compared against Critical Level values, can be used to evaluate the plant nutrient status at the time of sampling with a relatively high degree of confidence and can be extrapolated to project nutrient status at harvest. Soil analysis is similar to tissue analysis but evaluates the potential supplying power of the soil instead of plant nutrient status. Plant analysis provides information as to what the plant needs, while soil analysis provides information about the status of the nutrient supply.

Spectral analysis of nutrient status is still in its infancy and is presently used primarily in the inventory of global resources and in specialized studies. Microscopic studies are most valuable in looking at the physiological aspects of nutrient stress rather than the evaluation of plant nutrient status on a whole plant or crop basis.

Symptom Descriptions

It is unusual to find any one leaf or even one plant that displays the full array of symptoms that are characteristic of a given deficiency. It is thus highly desirable to know how individual symptoms look, for it is possible for them to occur in many possible combinations on a single plant. Most of the terms used below in the description of deficiency symptoms are reasonably self evident; a few however have a distinct meaning in the nutrient deficiency field. For example, the term *chlorotic,* which is a general term for yellowing of leaves through the loss of chlorophyll, cannot be used without further qualification because there may be an overall chlorosis as in nitrogen deficiency, interveinal, as in iron deficiency, or marginal, as in calcium deficiency. Another term used frequently in the description of deficiency symptoms is *necrotic,* a general term for brown, dead tissue. This symptom can also appear in many varied forms, as is the case with chlorotic symptoms.

Nutrient deficiency symptoms for many plants are similar, but because of the large diversity found in plants and their environments there is a range of expression of symptoms. Because of their parallel veins, grasses and other monocots generally display the affects of chlorosis as a series of stripes rather than the netted interveinal chlorosis commonly found in dicots. The other major difference is that the marginal necrosis or chlorosis found in dicots is often expressed as tip burn in monocots.

Magnesium: The Mg-deficient leaves show advanced interveinal chlorosis, with necrosis developing in the highly chlorotic tissue. In its advanced form, magnesium deficiency may superficially resemble potassium deficiency. In the case of magnesium deficiency the symptoms generally start with mottled chlorotic areas developing in the interveinal tissue.

The interveinal laminae tissue tends to expand proportionately more than the other leaf tissues, producing a raised puckered surface, with the top of the puckers progressively going from chlorotic to necrotic tissue. In some plants such as the Brassica (The mustard family, which includes vegetables such as broccoli, brussel sprouts, cabbage, cauliflower, collards, kale, kohlrabi, mustard, rape, rutabaga and turnip.), tints of orange, yellow, and purple may also develop.

Figure: Magnesium deficiency symptoms in tomato.

Manganese: These leaves show a light interveinal chlorosis developed under a limited supply of Mn. The early stages of the chlorosis induced by manganese deficiency are somewhat similar to iron deficiency. They begin with a light chlorosis of the young leaves and netted veins of the mature leaves especially when they are viewed through transmitted light. As the stress increases, the leaves take on a gray metallic sheen and develop dark freckled and necrotic areas along the veins. A purplish luster may also develop on the upper surface of the leaves. Grains such as oats, wheat, and barley are extremely susceptible to manganese deficiency. They develop a light chlorosis along with gray specks which elongate and coalesce, and eventually the entire leaf withers and dies.

Figure: *Manganese deficiency symptoms in tomato.*

Molybdenum: These leaves show some mottled spotting along with some interveinal chlorosis. An early symptom for molybdenum deficiency is a general overall chlorosis, similar to the symptom for nitrogen deficiency but generally without the reddish coloration on the undersides of the leaves. This results from the requirement for molybdenum in the reduction of nitrate, which needs to be reduced prior to its assimilation by the plant.

Thus, the initial symptoms of molybdenum deficiency are in fact those of nitrogen deficiency. However, molybdenum has other metabolic functions within the plant, and hence there are deficiency symptoms even when reduced nitrogen is available. In the case of cauliflower, the lamina of the new leaves fail to develop, resulting in a characteristic whiptail appearance. In many plants there is an upward cupping of the leaves and mottled spots developing into large interveinal chlorotic areas under severe deficiency. At high concentrations, molybdenum has a very distinctive toxicity symptom in that the leaves turn a very brilliant orange.

Figure: Molybdenum deficiency symptoms in tomato.

Nitrogen: The chlorotic symptoms shown by this leaf resulted from nitrogen deficiency. A light red cast can also be seen on the veins and petioles. Under nitrogen deficiency, the older mature leaves gradually change from their normal characteristic green appearance to a much paler green. As the deficiency progresses these older leaves become uniformly yellow (chlorotic).

Leaves approach a yellowish white colour under extreme deficiency. The young leaves at the top of the plant maintain a green but paler colour and tend to become smaller in size. Branching is reduced in nitrogen deficient plants resulting in short, spindly plants. The yellowing in nitrogen deficiency is uniform over the entire leaf including the veins.

However in some instances, an interveinal necrosis replaces the chlorosis commonly found in many plants. In some plants the underside of the leaves and/or the petioles and midribs develop traces of a reddish or purple colour.

In some plants this coloration can be quite bright. As the deficiency progresses, the older leaves also show more of a tendency to wilt under mild water stress and become senescent much earlier than usual. Recovery of deficient plants to applied nitrogen is immediate (days) and spectacular.

Figure: *Nitrogen deficiency symptoms in tomato.*

Phosphorus: These phosphorus-deficient leaves show some necrotic spots. As a rule, phosphorus deficiency symptoms are not very distinct and thus difficult to identify. A major visual symptom is that the plants are dwarfed or stunted. Phosphorus deficient plants develop very slowly in relation to other plants growing under similar environmental conditions but without phosphorus deficiency. Phosphorus deficient plants are often mistaken for unstressed but much younger plants. Some species such as tomato, lettuce, corn and the brassicas develop a distinct purpling of the stem, petiole and the under sides of the leaves. Under severe deficiency conditions there is also a tendency for leaves to develop a blue-gray luster. In older leaves under very severe deficiency conditions a brown netted veining of the leaves may develop.

Figure: *Phosphorus deficiency symptoms in tomato.*

Sulfur: This leaf shows a general overall chlorosis while still retaining some green colour. The veins and petioles show a very distinct reddish colour. The visual symptoms of sulfur deficiency are very similar to the chlorosis found in nitrogen deficiency. However, in sulfur deficiency the yellowing is much more uniform over the entire plant including young leaves. The reddish colour often found on the underside of the leaves and the petioles has a more pinkish tone and is much less vivid than that found in nitrogen deficiency. With advanced sulfur deficiency brown lesions and/or necrotic spots often develop along the petiole, and the leaves tend to become more erect and often twisted and brittle.

Figure: *Sulfur deficiency symptoms in tomato.*

Zinc: This leaf shows an advanced case of interveinal necrosis. In the early stages of zinc deficiency the younger leaves become yellow and pitting develops in the interveinal upper surfaces of the mature leaves. Guttation is also prevalent. As the deficiency progress these symptoms develop into an intense interveinal necrosis but the main veins remain green, as in the symptoms of recovering iron deficiency. In many plants, especially trees, the leaves become very small and the internodes shorten, producing a rosette like appearance.

Figure: *Zinc deficiency symptoms in tomato.*

Boron: These boron-deficient leaves show a light general chlorosis. The tolerance of plants to boron varies greatly, to the extent that the boron concentrations necessary for the growth of plants having a high boron requirement may be toxic to plants sensitive to boron. Boron is poorly transported in the phloem of most plants, with the exception of those plants that utilize complex sugars, such as sorbitol, as transport metabolites. In a recent study, tobacco plants engineered to synthesize sorbitol were shown to have increased boron mobility, and to better tolerate boron deficiency in the soil.

In plants with poor boron mobility, boron deficiency results in necrosis of meristematic tissues in the growing region, leading to loss of apical dominance and the development of a rosette condition. These deficiency symptoms are similar to those caused by calcium deficiency. In plants in which boron is readily transported in the phloem, the deficiency symptoms localize in the mature tissues, similar to those of nitrogen and potassium. Both the pith and the epidermis of stems may be affected, often resulting in hollow or roughened stems along with necrotic spots on the fruit. The leaf blades develop a pronounced crinkling and there is a darkening and crackling of the petioles often with exudation of syrupy material from the leaf blade. The leaves are unusually brittle and tend to break easily. Also, there is often a wilting of the younger leaves even under an adequate water supply, pointing to a disruption of water transport caused by boron deficiency.

Calcium: These calcium-deficient leaves show necrosis around the base of the leaves. The very low mobility of calcium is a major factor determining the expression of calcium deficiency symptoms in plants. Classic symptoms of calcium deficiency include blossom-end rot of tomato (burning of the end part of tomato fruits), tip burn of lettuce, blackheart of celery and death of the growing regions in many plants. All these symptoms show soft dead necrotic tissue at rapidly growing areas, which is generally related to poor translocation of calcium to the tissue rather than a low external supply of calcium. Very slow growing plants with a deficient supply of calcium may re-translocate sufficient calcium from older leaves to maintain growth with only a marginal chlorosis of the leaves. This ultimately results in the margins of the leaves growing more slowly than the rest of the leaf, causing the leaf to cup downward. This symptom often progresses to the point where the petioles develop but the leaves do not, leaving only a dark bit of necrotic tissue at the top of each petiole. Plants under chronic calcium deficiency have a much greater tendency to wilt than non-stressed plants.

Chloride: These leaves have abnormal shapes, with distinct interveinal chlorosis. Plants require relatively high chlorine concentration in their tissues. Chlorine is very abundant in soils, and reaches high concentrations in saline areas, but it can be deficient in highly leached inland areas. The most common symptoms of chlorine deficiency are chlorosis and wilting of the young leaves. The chlorosis occurs on smooth flat depressions in the interveinal area of the leaf blade. In more advanced cases there often appears a characteristic bronzing on the upper side of the mature leaves. Plants are generally tolerant of chloride, but some species such as avocados, stone fruits, and grapevines are sensitive to chlorine and can show toxicity even at low chloride concentrations in the soil.

Copper: These copper-deficient leaves are curled, and their petioles bend downward. Copper deficiency may be expressed as a light overall chlorosis along with the permanent loss of turgor in the young leaves. Recently matured leaves show netted, green veining with areas bleaching to a whitish gray. Some leaves develop sunken necrotic spots and have a tendency to bend downward. Trees under chronic copper deficiency develop a rosette form of growth. Leaves are small and chlorotic with spotty necrosis.

Iron: These iron-deficient leaves show strong chlorosis at the base of the leaves with some green netting. The most common symptom for iron deficiency starts out as an interveinal chlorosis of the youngest leaves, evolves into an overall chlorosis, and ends as a totally bleached leaf. The bleached areas often develop necrotic spots. Up until the time the leaves become almost completely white they will recover upon application of iron. In the recovery phase the veins are the first to recover as indicated by their bright green colour. This distinct venial re-greening observed during iron recovery is probably the most recognizable symptom in all of classical plant nutrition. Because iron has a low mobility, iron deficiency symptoms appear first on the youngest leaves. Iron deficiency is strongly associated with calcareous soils and anaerobic conditions, and it is often induced by an excess of heavy metals.

Potassium: Some of these leaves show marginal necrosis (tip burn), others at a more advanced deficiency status show necrosis in the interveinal spaces between the main veins along with interveinal chlorosis. This group of symptoms is very characteristic of K deficiency symptoms. The onset of potassium deficiency is generally characterized by a marginal chlorosis progressing into a dry leathery tan scorch on recently matured leaves. This is followed by increasing interveinal scorching and/or necrosis progressing from the leaf edge to the midrib as the stress increases. As the deficiency progresses, most of the interveinal area becomes necrotic, the veins remain green and the leaves tend to curl and crinkle. In some plant such as legumes and potato, the initial symptom of deficiency is white speckling or freckling of the leaf blades. In contrast to nitrogen deficiency, chlorosis is irreversible in potassium deficiency, even if potassium is given to the plants. Because potassium is very mobile within the plant, symptoms only develop on young leaves in the case of extreme deficiency. Potassium deficiency can be greatly alleviated in the presence of sodium but the resulting sodium-rich plants are much more succulent than a high potassium plant. In some plants over 90% of the required potassium can be replaced with sodium without any reduction in growth.

3

Choice of Material in Physiological Test

Although it is possible to measure respiration by intact tissues it is often useful to be able to isolate uncontaminated, intact and functional mitochondria. Here, we will consider some of the principles for isolating cell organelles. This discussion is not meant to be exhaustive nor will it be a cookbook with recipes to be followed step-by-step.

If the aim of the experiment is to investigate a specific physiological problem, then the plant material is often well-defined. However, in many cases there is some flexibility in the choice of material and it is therefore wise to follow certain rules of thumb:

- Choose a tissue that is easy to obtain in relatively large amounts (at least 100–200 g fresh weight), such as tubers or hypocotyls.
- Choose a tissue for which a good protocol is available. It can take quite a lot of time and effort to optimize an isolation protocol.

Note that the isolation of mitochondria from green tissues presents special problems because of the large amount of chloroplasts. Roots are a notoriously difficult tissue for isolation of mitochondria.

Disruption of the Tissue

The aim is to rupture as many of the cells in the tissue as possible in order to release their cytoplasm with the mitochondria. Whereas many mammalian tissues are easy to homogenize, plant tissues are much tougher mainly because of the cell walls. There are a number of standard homogenizers, including mixers with rotating knives or razor blades, juice extractors, ball mills, and mortar and pestle; the most suitable method for the specific plant material should be chosen. This may involve some initial optimization experiments.

Homogenization Medium

The medium for homogenizing rat liver only contains an osmoticum and a buffer. That will not work with plant tissues because of the disruption of the large central vacuole (absent in mammalian cells), which releases its fairly acidic content often containing phenols and other potentially harmful substances.

Let us consider the components of the optimal homogenization medium for isolating intact and functional organelles from plant material, one by one.

Osmoticum. This is required to prevent the mitochondria from losing their matrix content because of swelling and bursting. For the isolation of plant mitochondria, there is a tradition of using sucrose or mannitol at 0.2–0.5 M; whereas 0.15 M KCl is standard for the isolation of mammalian mitochondria. In neither case does this mimic precisely the conditions in the cytosol, but it works.

Buffer. The pH in the cytosol of plant cells is around 7.5, and the homogenate should be buffered at pH 7–8. A buffer should therefore be included and a plethora of different buffers are available. Choose an inexpensive one with a pKa of 7–8 (e.g., 4-morpholinopropane sulfonic acid = MOPS) and be sure to use enough to counterbalance the acidic vacuolar content. The tissue type, the tissue/medium ratio, and the buffer concentration determine whether the acidic vacuolar content is properly neutralized. Obviously, special tissues like lemons or CAM plants harvested late at night or in the early morning, require special treatment!

Metal ion chelators. EDTA is usually included to bind mainly Ca^{2+} and Mg^{2+}, which may otherwise activate proteolytic enzymes.

Antioxidants. Antioxidants (reductants) such as ascorbic acid, dithiothreitol, or cysteine are used mainly to protect sulfhydryl groups on proteins against oxidation.

Bovine serum albumin (BSA). This protein from the blood of cows is usually included to protect the mitochondria: BSA binds fatty acids (its function in the blood) released by the action of phospholipases on phospholipids. BSA may also bind phenolics that are released from the vacuole. Both of these groups of compounds can otherwise be troublesome later. In addition, BSA may prevent proteases from degrading mitochondrial proteins by serving as an alternative substrate present at high concentrations.

Polyvinylpyrrolidone (PVP). A polymer that is often included, especially when using leaf material, to adsorb phenolic compounds.

Filtration

To remove larger tissue fragments, cell walls, and often starch grains, the homogenate is always filtered, usually through a nylon net with a mesh size of 30–100 μm.

Purification of Mitochondria

The first step of the purification is usually differential centrifugation. This means that the filtered homogenate is centrifuged first at low speed for a short time (e.g., 3000 g for 5–10 min.) and the pellet (containing cell wall fragments, starch grains, nuclei, intact plastids) is discarded. The supernatant is then centrifuged faster and for a longer time (e.g., 10,000 g for 10–20 min.) and the pellet (termed *crude mitochondria*) is collected. It contains intact and damaged mitochondria as well as peroxisomes and plastid membranes. The supernatant is discarded.

As soon as the mitochondria have been pelleted and thus removed from potentially damaging soluble substances in the homogenate, the medium can be greatly simplified. Usually the so-called wash medium contains only osmoticum and weak buffer (no need for a strong buffer since there is no vacuole content to neutralize). However, it may be wise to include EDTA to bind divalent cations as well as BSA to adsorb any fatty acids released by the action of phospholipases. The resuspended crude mitochondria are often diluted to 50–100 ml with wash medium and subjected to another round of slow and fast centrifugation to remove more contaminants.

The aim of the next step is to separate the intact mitochondria from the contaminants. This usually involves a density gradient centrifugation with Percoll either alone (e.g., Struglics et al. 1993) or together with other dense materials (e.g., Day et al. 1985). Percoll (colloidal silica coated with PVP) is inert and osmotically inactive. It has the further advantage that a Percoll gradient is self-generating because the Percoll microspheres pellet very slowly during the centrifugation. The crude mitochondria are layered on top of the wash medium containing the appropriate concentration of Percoll (usually 25–30%). If the Percoll concentration and the speed and duration of the centrifugation have been chosen correctly, the intact mitochondria will form a distinct band, well away from the contaminants. This band can be collected, washed free of Percoll by repeated dilution and centrifugation, and either used immediately, or frozen with 5% (v/v) dimethylsulfoxide (to preserve integrity) in liquid nitrogen and stored at -80°C until use. This preparation can be stable for months, retaining the ability to form a proton electrochemical gradient over the inner membrane.

The yield is 1–50 mg mitochondrial protein per 100 g fresh weight, depending on the tissue. This represents only a small fraction of the total amount of mitochondria in the tissue. The purification of mitochondria from green leaves presents special problems and a good protocol is described in Day et al. (1985). There is, at present, no good protocol for the isolation of both pure and functional mitochondria from Arabidopsis leaves (Hausmann et al. 2003; Keech et al. 2005), but a protocol for the purification of mitochondria from Arabidopsis cell cultures is given in Kruft et al. (2001). Finally, Boutry et al. (1984) demonstrate how purified mitochondria can be isolated from small amounts of plant tissue (down to 0.5 g fresh weight).

Mitochondria are rod-shaped organelles that can be considered the power generators of the cell, converting oxygen and nutrients into adenosine triphosphate (ATP). ATP is the chemical energy "currency" of the cell that powers the cell's metabolic activities.

This process is called aerobic respiration and is the reason animals breathe oxygen. Without mitochondria (singular, mitochondrion), higher animals would likely not exist because their cells would only be able to obtain energy from anaerobic respiration (in the absence of oxygen), a process much less efficient than aerobic respiration. In fact, mitochondria enable cells to produce 15 times more ATP than they could otherwise, and complex animals, like humans, need large amounts of energy in order to survive.

The number of mitochondria present in a cell depends upon the metabolic requirements of that cell, and may range from a single large mitochondrion to thousands of the organelles. Mitochondria, which are found in nearly all eukaryotes, including plants, animals, fungi, and protists, are large enough to be observed with a light microscope and were first discovered in the 1800s. The name of the organelles was coined to reflect the way they looked to the first scientists to observe them, stemming from the Greek words for "thread" and "granule." For many years after their discovery, mitochondria were commonly believed to transmit hereditary information. It was not until the mid-1950s when a method for isolating the organelles intact was developed that the modern understanding of mitochondrial function was worked out.

The elaborate structure of a mitochondrion is very important to the functioning of the. Two specialized membranes encircle each mitochondrion present in a cell, dividing the organelle into a narrow intermembrane space and a much larger internal matrix, each of which contains highly specialized proteins. The outer membrane of a mitochondrion contains many channels formed by the protein porin and acts like a sieve, filtering out molecules that are too big. Similarly, the inner membrane, which is highly convoluted so that a large number of infoldings called cristae are formed, also allows only certain molecules to pass through it and is much more selective than the outer membrane. To make certain that only those materials essential to the matrix are allowed into it, the inner membrane utilizes a group of transport proteins that will only transport the correct molecules. Together, the various compartments of a mitochondrion are able to work in harmony to generate ATP in a complex multi-step process.

Mitochondria are generally oblong organelles, which range in size between 1 and 10 micrometers in length, and occur in numbers that directly correlate with the cell's level of metabolic activity. The organelles are quite flexible, however, and time-lapse studies of living cells have demonstrated that mitochondria change shape rapidly and move about in the cell almost constantly. Movements of the organelles appear to be linked in some way to the microtubules present in the cell, and are probably transported along the network with motor proteins. Consequently, mitochondria may be organized into lengthy travelling chains, packed tightly into relatively stable groups, or appear in many other formations based upon the particular needs of the cell and the characteristics of its microtubular network.

The mitochondrion is different from most other organelles because it has its own circular DNA (similar to the DNA of prokaryotes) and reproduces independently of the cell in which it is found; an apparent case of endosymbiosis. Scientists hypothesize that millions of years ago small, free-living prokaryotes were engulfed, but not consumed, by larger prokaryotes, perhaps because they were able to resist the digestive enzymes of the host organism. The two organisms developed a symbiotic relationship over time, the larger organism providing the smaller with ample nutrients and the smaller organism providing ATP molecules to the larger one. Eventually, according to this view, the larger organism developed into the eukaryotic cell and the smaller organism into the mitochondrion.

Mitochondrial DNA is localized to the matrix, which also contains a host of enzymes, as well as ribosomes for protein synthesis. Many of the critical metabolic steps of cellular respiration are catalysed by enzymes that are able to diffuse through the mitochondrial matrix. The other proteins involved in respiration, including the enzyme that generates ATP, are embedded within the mitochondrial inner membrane. Infolding of the cristae dramatically increases the surface area available for hosting the enzymes responsible for cellular respiration.

Methods for Establishing Purity, Intactness, and Functionality of the Fractions

It is often important to determine the degree of purity of the mitochondria and their subfractions.

The intactness of the mitochondria is usually established with enzyme assays that take advantage of the fact that an intact membrane is not permeable to certain molecules.

For instance, the outer mitochondrial membrane is impermeable to cytochrome *c*, a 12.5 kDa protein that is too large to pass through the outer membrane pores, and NADH cannot pass through an intact inner membrane.

Functionality is usually tested using an oxygen electrode. If the mitochondria show high rates of oxygen consumption, high respiratory control ratios (>3), and ADP:O ratios close to the theoretical values they are in very good shape. Note that, for mitochondria with an appreciable alternative oxidase activity (measured in the presence of the cytochrome oxidase inhibitor, KCN), the respiratory control ratio and ADP:O ratio should be measured in the presence of an alternative oxidase inhibitor such as salicylhydroxamic acid.

Mitochondrial Activities can be Determined in Situ

It is relatively straightforward to measure the activity of enzymes in the soluble matrix fraction isolated from mitochondria.

However, this fraction is very diluted compared to the matrix inside intact mitochondria (300–500 mg protein/ml!) and protein–protein interactions may have been lost. When measuring the activity of enzymes in fractions containing membranes—intact mitochondria and submitochondrial particles—a detergent can often be used to remove the permeability barrier that the inner membrane constitutes.

However, it is also possible to use the ionophore alamethicin to permeabilize the inner membrane to molecules <1000 Da, which makes it possible to measure the activity of matrix and inner membrane enzymes in their native proteinaceous environment.

Using alamethicin, it is even possible to measure the activity of cytosolic and mitochondrial enzymes inside intact cells since the ionophore also permeabilizes the plasma membrane.

The Q-Cycle Explains How Protons Can Be Pumped across the Inner Mitochondrial Membrane

The electron transport chain (ETC) in the mitochondrial inner membrane of most eukaryotes consists of four complexes, complex I to complex IV. Complexes I, III and IV pump protons over the inner mitochondrial membrane when transporting electrons.

For each electron pair (corresponding to one NADH or one oxygen atom), four protons are pumped by complex I, four by complex III, and two by complex IV. The detailed three-dimensional structures have been determined for the mammalian complexes III and IV, but not for complex I. The structural information has improved the understanding of the electron paths of complexes III and IV, but the proton-pumping function of the latter is still not fully understood. For complex I, neither electron nor proton paths are known in any detail.

Ubiquinone (Q), also called coenzyme Q, is a mobile electron and proton carrier in the inner mitochondrial membrane, structurally and functionally similar to plastoquinone in the thylakoid membrane of plastids (Figure 1). The ability of Q to carry both electrons and protons—and that it can accept and donate them step-wise—is utilized in electron transport and proton pumping (Figure 2) (Nicholls and Ferguson 2002).

To be reduced, it can receive two electrons from complex I or succinate dehydrogenase and take up two protons from the matrix, forming QH_2. However, at the oxidizing side (complex III), there is actually both oxidation and reduction taking place. A QH_2 will be oxidized to Q at the outer Q-site, releasing two electrons. One of these electrons goes to cytochrome *c*, whereas the other is recycled to reduce an oxidized Q in the inner Q-site—one step to becoming a semiquinone.

The next time a QH_2 is oxidized at the outer Q-site, the semiquinone in the inner site will receive a second electron, becoming a QH_2. So, from two QH_2, we have made one Q and one QH_2, which again can be oxidized. In each cycle (i.e., for each electron passing to cytochrome *c*), two protons are taken up from the matrix and two protons released to the intermembrane space. So, the Q-cycle mediates the pumping of an extra proton per electron transported, increasing energy conservation.

If we look at the whole reaction from succinate to cytochrome *c*, we see that for each succinate, four protons are transported from the matrix to the intermembrane space.

However, at the same time, the two electrons are also transported out. So, this corresponds to the net pumping out of two positively charged species. Since, energetically, the mitochondrial membrane potential (which determines the energetic cost of transporting a positive charge out) is much larger than the pH gradient (which determines the cost for electroneutral export of a proton), the complex III proton pumping is less energy-demanding than, for example, complex I, which pumps four protons in their completeness.

Multiple Energy-Conservation Bypasses in Oxidative Phosphorylation of Plant Mitochondria

The electron transport chain (ETC) in the mitochondrial inner membrane of most eukaryotes consists of four large protein complexes, complex I to complex IV. The electron transport activity of complexes I, III and IV (the latter two, often jointly called the cytochrome pathway) is coupled to extrusion of protons across the inner mitochondria.

The electrochemical proton gradient that is formed and maintained in the process is used by the ATP synthase (also called complex V) to make ATP (Figure 1). The plant ETC, however, is highly branched. As a consequence, the plant oxidative phosphorylation system has several alternative pathways of electron transport via type II NAD(P)H dehydrogenases and the alternative oxidase (AOX), and additionally, an alternative proton transporter called the uncoupling protein (UCP).

These proteins mediate bypasses around the proton-translocating multi-protein complexes (I–V): type II NAD(P)H dehydrogenases bypass complex I, AOX bypasses complexes III and IV, and UCP bypasses the ATP synthase (Siedow and Umbach 1995; Vanlerberghe and McIntosh 1997; Møller 2001; Vercesi 2001; Millenaar and Lambers 2003; Fernie et al. 2004; Rasmusson et al. 2004).

The type II NAD(P)H dehydrogenases and the AOX transport electrons without pumping protons, whereas the UCP allows proton flow from the intermembrane space to the matrix without ATP synthesis. The consequence of this is that flux through the energy conservation bypasses will neither contribute to respiratory ATP production, nor will they be controlled by cellular adenylate status. The special features of these energy bypass proteins and discuss the consequences of their existence to the plant.

NAD(P)H Dehydrogenases in Plant Mitochondria

Except for yeast, the ETC of all eukaryotes—including plants—contains a proton-pumping NADH dehydrogenase, complex I, which is inhibited specifically by rotenone. Additionally, plant, fungal and protist mitochondria contain different setups of rotenone-insensitive enzymes capable of oxidizing NADH and/or NAD(P)H and passing the electrons on to ubiquinone.

These NAD(P)H dehydrogenases are classified as type II, whereas complex I belongs to type I. The inner mitochondrial membrane in plants contains such NAD(P)H dehydrogenases, facing either the matrix or the intermembrane space, thus oxidizing NAD(P)H either from the matrix or the cytoplasm.

The outer membrane of all mitochondria contains NADH-ferricyanide and NADH-cytochrome *b* reductase activities that are insensitive to rotenone and to the complex III inhibitor, antimycin A.

There are Three NAD(P)H Dehydrogenases on the Inner Matrix Surface of the Inner Membrane

Much of the information about the number of matrix-facing NAD(P)H dehydrogenases has come from experiments using inside-out inner membrane vesicles. Such vesicles have the matrix surface facing the medium and the external NAD(P)H dehydrogenases hidden inside the vesicles and unable to interfere with the assays.

Recently, this method has been complemented with a technique where one uses a peptide called alamethicin to make a channel for the substrate through the inner membrane. Experiments with these methods show the presence of three NAD(P)H dehydrogenase activities on the matrix side of the inner membrane:

- Complex I pumps protons, is Ca^{2+}-independent and is inhibited by rotenone and diphenylene iodonium (DPI). Complex I has a high affinity for NADH.
- ND_{in}(NADH) does not pump protons, is Ca^{2+}-independent and DPI insensitive. The enzyme has a tenfold lower affinity for NADH than complex I.
- ND_{in}(NADPH) does not pump protons, is Ca^{2+}-dependent and very sensitive to DPI.

Because of the contrasting affinities of complex I and ND_{in}(NADH) for NADH, only complex I is engaged at low matrix concentrations of NADH. However, when the NADH concentration in the matrix increases (e.g., if the ETC is restricted by lack of ADP in the cell), ND_{in}(NADH) can become engaged. When electrons are transported through the type II NAD(P)H dehydrogenases, only the two proton pumping sites of the cytochrome pathway are involved in the flow of electrons from matrix NADH to oxygen. As a consequence, the theoretical ADP/O yield is 1.5 unless the AOX or the UCP are active, in which case it will be even lower.

The External NAD(P)H Dehydrogenases are Ca^{2+}-Dependent

Plant mitochondria can oxidize cytosolic NADH and NAD(P)H directly via external dehydrogenases. The NADH and NAD(P)H dehydrogenase activities are catalyzed by separate enzymes. None of the external dehydrogenases pumps protons, which means that, at most, two sites of proton-pumping are involved when the electrons move from the dehydrogenases to oxygen. So, as for ND_{in}(NADH), the theoretical maximum ADP/O yield is 1.5.

The oxidation of exogenous NADH or NAD(P)H by most plant mitochondria requires Ca^{2+}, and as a result, is inhibited by EDTA and EGTA, chelators of di-and trivalent cations. *In vitro*, the Ca^{2+}concentration needed for half-maximal activity of NADH oxidation is 0.1–0.5 μM, depending on ionic environment and respiratory state This is in the same range as the 0.1–0.2 μM free Ca^{2+} thought to be present in the cytosol of a resting plant cell. Therefore, the external NAD(P)H oxidation may become activated by the increased Ca^{2+}-concentration associated with cellular excitation due to external conditions.

Mammalian mitochondria lack external NAD(P)H dehydrogenases. Instead, they have a Ca^{2+}-dependent glycerol-3-phosphate dehydrogenase on the outer surface of the inner membrane, which donates electrons directly to the ubiquinone pool.

The dihydroxyacetone phosphate formed is converted back into glycerol-3-phosphate by a cytosolic NAD-dependent glycerol-

3-phosphate dehydrogenase. In this way, the two low-molecular-weight compounds, glycerol-3-phosphate and dihydroxyacetone phosphate, shuttle redox equivalents from the cytosol, across the outer membrane and to the inner membrane to yield the same amount of ATP as is produced by the plant mechanism. Yeast have both external NADH dehydrogenases and a glycerol-3-phosphate shuttle, and since a mitochondrial glycerol-3-phosphate dehydrogenase has been found in plant mitochondria, plants may also have both systems.

Another way to oxidize cytosolic NAD(P)H that is found in mammalian mitochondria is to use the malate/aspartate shuttle to tranfer reductant (a collective term for high-energy electrons carried by, for example, NAD(P)H, ferredoxin or an organic acid redox couple like malate/oxaloacetate) into the matrix as NADH that can be oxidized by the internal NAD(P)H dehydrogenases. However, it is not clear whether this shuttle can work for import of cytosolic reductant into plant mitochondria.

In the opposite orientation, NADH is shuttled from the matrix to the cytosol via the malate/oxaloacetate shuttle. In this way, the external NADH dehydrogenase can also work as a bypass around complex I. Support for this bypass has been seen in the form of an upregulation of external NADH dehydrogenase activity in mutants where complex I or an internal type II NAD(P)H dehydrogenase has been inactivated.

The Molecular Identities of the Rotenone-Insensitive NAD(P)H Dehydrogenases

The type II NADH dehydrogenases from *E. coli* and yeast are small, single polypeptide proteins of around 50 kDa. They oxidize only NADH and contain FAD as an electron-accepting prosthetic group.

Attempts to purify the type II NAD(P)H dehydrogenases have been unsuccessful in identifying the plant proteins. However, the characterization of potato homologues to the type II NADH dehydrogenases in *E. coli* and yeasts have provided candidate plant proteins to try and assign to the different reported activities.

One protein, NDA1, is internal and the other, NDB1, external of the inner mitochondrial membrane. The latter was shown to be an external calcium-dependent NAD(P)H dehydrogenase. Like a related fungal NAD(P)H dehydrogenase, NDB1 contains a motif for Ca-binding, indicating that in the activation of plant and fungal NAD(P)H dehydrogenase by Ca^{2+}, the ion binds directly to the enzyme.

In Arabidopsis, a weed for which the nuclear genome sequence is known, small families of both NDA-type and NDB-type proteins are encoded by nuclear genes. One of the NDA-type proteins has been shown to be an internal NAD(P)H dehydrogenase. Additionally, a third type of type II NAD(P)H dehydrogenase was found in Arabidopsis, the NDC1. This protein is most closely related to cyanobacterial enzymes, suggesting that the gene entered the eukaryotic domain with the plastid, later to be transferred to the nucleus.

The Alternative Oxidase

The presence of the AOX is one of the features that sets plant (and fungal) mitochondria apart from mammalian mitochondria. This enzyme has therefore received a lot of attention from plant scientists and we now have a fairly good understanding of the regulation of its structure and activity.

The AOX is a quinol-oxygen oxidoreductase and it does not pump protons. The oxygenase transfers electrons from ubiquinol (the reduced form of ubiquinone) to oxygen and generates water as the end-product of the reaction, so four electrons must be transferred to oxygen.

The precise mechanism is not known, but the protein has been shown to contain a catalytical di-iron centre. The functional form of the enzyme is a dimer with the two polypeptides either covalently or non-covalently bound to each other (Figure 2). The predicted structure places the protein embedded into the inner leaflet of the inner mitochondrial membrane, with the active site and the regulatory cysteines exposed to the matrix.

When electrons from NADH oxidation flow through the AOX, at least two of the three sites of proton translocation are bypassed, so energy conservation in the form of ATP is much smaller when the AOX is active. The activity of the AOX must therefore be carefully regulated. There are often several AOX isoforms in each plant species—for example, five genes encoding AOX proteins are present in Arabidopsis. The relative amounts of three isoforms of AOX vary with development, and cellular redox status, as well as external factors.

The AOX is present in the inner membrane as a dimer in which the two subunits can be covalently linked through a disulfide bridge between cysteine residues. The reduced non-covalently linked dimer is the active form, and this form appears to be the predominant one *in vivo*. It was recently shown that a mitochondrial thioredoxin system regulates AOX. Thioredoxins regulate enzymes by reducing disulfide bridges, and it is likely that the regulation involves the redox-active cysteines.

The presence of thioredoxin regulation connects the AOX activation level to the mitochondrial NAD(P)H pool and the oxidation of citric acid cycle metabolites by NADP-utilizing enzymes.

The activity of AOX is also stimulated by 2-oxo organic acids—in particular, pyruvate, which activates by interacting with the same cysteine that is involved in dimerization. This activation may be a feed-forward mechanism that upregulates the capacity of the ETC when the glycolytic end-product is available. Both the cytochrome pathway and the AOX pathway use ubiquinol as the substrate. In the absence of pyruvate, the cytochrome pathway is saturated at 40% Q_{red}/Q_{tot} (40% of the UQ is reduced), at which point the AOX pathway is not engaged at all. The AOX pathway starts accepting electrons at 50% reduction and is not fully saturated at 80%.

In the presence of pyruvate, the AOX becomes engaged at a much lower Q_{red}/Q_{tot} (20%), whereas the cytochrome pathway is unaffected. It is puzzling, however, that the normal *in vivo* concentration of pyruvate appears to be higher than the concentration of pyruvate required to give maximal stimulation of AOX activity *in vitro*. It thus appears that there is little regulation of AOX activity via pyruvate under physiological conditions.

AOX Activity Can Be Measured In Vivo

For many years, benzhydroxamic acid derivatives,, were used to quantify the capacity and activity of the AOX pathway in both isolated mitochondria and intact tissues. AOX capacity was determined from the rate of respiration in the presence of KCN (which blocks the cytochrome pathway), and corrected for residual respiration in the presence of both KCN and SHAM.

AOX *in vivo* activity was determined from the decrease in the rate of respiration when SHAM was added in the absence of KCN. The assumption was that the cytochrome pathway was always used first and that the AOX would be engaged only when the cytochrome pathway was saturated. However, as we have seen above, this assumption is not correct. The consequence of this is that *in vivo* activity of AOX cannot be measured using inhibitors alone.

A better method for measuring AOX *in vivo* activity is now available. It makes use of the fact that cytochrome *c* oxidase and AOX discriminates differently between ^{18}O relative to ^{16}O.

The so-called oxygen discrimination of the sample (tissue or mitochondria) can be measured by mass spectrometry: (1) with KCN to resolve the oxygen discrimination by AOX; (2) with SHAM to resolve the oxygen discrimination by cytochrome *c* oxidase; and (3) without inhibitors to resolve the discrimination value resulting from the activity of both enzymes.

From these values, the percentage contribution by AOX and cytochrome *c* oxidase to the total uninhibited rate of respiration can be calculated. An example of the use of the oxygen discrimination technique with an intact tissue is given below. A remaining problem in the estimation of AOX activity *in vivo* is that measurements can only be made in darkness (to avoid interference from photosynthetic oxygen exchange).

Plant Mitochondria also Contain an Uncoupling Protein

Mammalian mitochondria contain neither type II NAD(P)H dehydrogenases nor AOX. However, they have a UCP that increases the proton permeability of the inner mitochondrial membrane and, in that way, dissipates the proton gradient. This is another mechanism for reducing the ATP production and increasing heat production, for example, for thermoregulation.

Surprisingly, plant mitochondria also contain a UCP. We do not know for certain why plant mitochondria should require two different mechanisms for achieving the same end result. The UCP is related to anion carriers. Its *in vitro* transport activity is dependent on free fatty acids and is inhibited by nucleotides like ATP and GTP but stimulated by superoxide. The latter result indicates that the UCP may protect against oxidative stress.

The Physiological Impact of Energy Bypass Proteins

The respiratory rates of some thermogenic flowers are the highest among plants and, in fact, exceed even those of warm-blooded animals. The external NADH dehydrogenase, AOX and UCP appear particularly active in thermogenic flowers where they are primarily responsible for heat production. At the same time it is clear that the respiratory rate of most plant tissues is insufficient to generate an appreciable increase in tissue temperature.

AOX is up-regulated by cold stress, although this is clearly not to heat the tissue. However, during cold stress, reactive oxygen species (ROS) are formed, and also other stress treatments that lead to elevated ROS will induce AOX. ROS (e.g., superoxide and hydrogen peroxide) are formed as by-products of electron transport under aerobic conditions.

ROS can cause damage to proteins, lipids, and DNA and the cell must therefore limit their formation. AOX helps prevent overreduction of the ETC and thus lowers ROS production. Supporting this, Maxwell et al. (1999) demonstrated the importance of the AOX for limiting ROS formation in living plant cells.

Overexpression of AOX lowered the steady state level of ROS, whereas suppression increased it. This result also points to the mitochondrion as an important site of ROS generation in plant cells. With the recent discovery of ROS stimulation of plant UCP and the increased ROS resistance in plants overexpressing UCP, the picture has become more complex. By decreasing the resistance for electron transport through the proton-pumping complexes, UCP activity may, like AOX, lead to decreased overreduction and ROS formation. Another factor speaking in favour of UCP having a role in ROS protection is that the AOX is inhibited by an ROS product.

Both type II NAD(P)H dehydrogenases and AOX may act as reductant overflow enzymes by reoxidizing surplus NADH to oxygen. This can prevent inhibition of the citric acid cycle under conditions in which the ADP concentration is low, and thus the proton-pumping complexes are restricted by a high membrane potential. In this way, the citric acid cycle can supply carbon skeletons for biosynthesis, even if the ATP level in the cell is high.

The concept of overflow is also relevant for photorespiratory conditions, under which there is a massive flow of glycine that has to be oxidized by the mitochondria. As much as half of the NADH produced in this process is probably exported, but the other half has to be oxidized promptly. There are several reasons to believe that a reductant overflow via energy bypass proteins is involved.

The ND_{in}(NADH) and AOX capacity in leaf mitochondria is higher than in mitochondria from heterotrophic tissue, and both proteins are down-regulated by dark treatment. Consistent with these observations, AOX activity *in vivo* is up-regulated by light (Ribas-Carbo et al. 2000), AOX capacity is induced by greening of etiolated tissue, and expression of the *nda*1 gene [encoding an internal NAD(P)H dehydrogenase] in potato and Arabidopsis is light-dependent.

Mutant tobacco plants, lacking the proton-pumping complex I can still survive, apparently aided by the action of ND_{in}(NADH), which oxidizes matrix NADH with a lower ATP yield.

These plants, however, show decreased photosynthetic rates, indicating that the ND_{in}(NADH) cannot fully compensate for the missing complex I. Also, external NAD(P)H dehydrogenases have been suggested to support photosynthesis. These enzymes may reoxidise NAD(P)H that is shuttled out of the chloroplast in order to balance the NAD(P)H to ATP ratio in the stroma. In summary, there are many indications that mitochondria—especially the energy bypass proteins—have roles in supporting photosynthesis, for example by oxidizing excess NAD(P)H.

The AOX pathway is involved in the acidification/deacidification cycle of CAM plants. CAM plants fix CO_2 in the form of malate during the night. This is done by cytosolic PEP carboxylase, and the malate produced is stored in the vacuole.

The CO_2 fixed is released during the daytime by malic enzymes, located in cytosol and mitochondria. Robinson et al. (1992) used the oxygen discrimination technique to measure the contribution of the AOX pathway to whole-leaf respiration in the CAM-plant, *Kalanchoë daigremontiana.*

They found that the respiratory rate of the leaves increased by 35% in the early hours of the day. This increase was due to increased AOX activity. A more detailed time course showed that the AOX activity increased two-to threefold during the early day at the same time as the malate was metabolized. As soon as the malate had disappeared by mid-day, the AOX activity returned to the low level observed during the night. A reasonable conclusion is that a substantial amount of the malate from the vacuole is oxidized via malic enzyme in the mitochondria. This generates NADH in the matrix, which is oxidized to a large extent via the AOX. The released CO_2 can then be refixed by the Calvin cycle in the chloroplast.

F_oF_1-ATP Synthases: The Worlds Smallest Rotary Motors

F_oF_1-ATP synthases (also called F-ATPases) are present in the inner membranes of mitochondria, chloroplasts, and bacteria. These large, multisubunit enzymes consist of a water-soluble catalytic complex (F_1) attached to an integral membrane protein complex (F_o) that transports protons across the membrane. The F_1 complex is composed of at least five different types of subunits (three α, three β, one γ, one δ, and one å).

When the F_1 complex is dissociated from the membrane, it is active as an ATPase. In fact, under the appropriate conditions the intact F_oF_1-ATP synthase can run in reverse and act as a proton pump, using the energy of ATP hydrolysis to move H^+ across the membrane. Our understanding of how ATP is synthesized was advanced by Paul Boyer at the University of California, Los Angeles, who proposed the binding-change mechanism for catalysis by F-ATPases. From the detailed understanding of how the ATPases function, he proposed that the binding-change mechanism for ATP synthesis contains three important components:

1. The major energy-requiring step is not the synthesis of ATP from ADP and P_i, but the release of ATP from the enzyme.

2. Substrate is bound and products are released at three separate but interacting catalytic sites, corresponding to the three catalytic subunits (b subunits). Each catalytic site can exist in one of three conformations: tight, loose, or open.
3. The binding changes are coupled to proton transport by rotation of the g subunit. That is, the flow of protons down their electrochemical gradient through the F_o complex causes the g subunit to rotate.

Rotation of the γ subunit then brings about the conformational changes in the catalytic complex that allow the release of ATP from the enzyme, and the reaction is driven forward. The reverse occurs when the enzyme functions as an ATP-driven proton pump.

The first two predictions of the binding-change model are supported by many lines of evidence, mainly kinetic studies, and are now generally accepted. However, the prediction of a rotary mechanism for coupling proton flow to ATP synthesis has been more difficult to demonstrate. Two major breakthroughs have led to confirmation of the third prediction, as well.

The first breakthrough was the determination of the crystal structure of the F_1-part of bovine mitochondrial ATPase by the laboratory of John Walker in Cambridge, England. The crystal structure showed that the three catalytic β subunits differ in their conformations and in the nucleotide bound to them, consistent with the binding-change mechanism.

Even more exciting was the discovery that the γ subunit is inserted like a shaft through the centre of the catalytic complex, which consists of three α subunits and three β subunits arranged alternately in a doughnutlike structure. Moreover, the interface between the γ subunit and the α and β subunits is highly hydrophobic.

The hydrophobicity of the interface minimizes the interactions between the subunits, consistent with the rotation of the γ subunit within the hole formed by the catalytic complex. In other words, the γ subunit looks like a molecular bearing lubricated by a hydrophobic interface. Although many questions were answered by elucidation of the crystal structure of the F_1-ATP synthase, the rotational model cannot be tested by a static "snapshot" of the enzyme. Definitive demonstration of rotation requires a video recording of the spinning of the enzyme in real time. But although they are large for proteins, F_1-ATP synthases are still far too small to be visualized in a light microscope.

To visualize the rotation of the enzyme, Masasuke Yoshida and his colleagues at the Tokyo Institute of Technology came up with an ingenious method to make the enzyme much larger than it is. As Figure below shows, they attached an actin filament labelled with a fluorescent dye to the base of the γ subunit using another protein as a "glue." They then attached the F_1 complex upside down to a glass surface.

If the γ subunit rotates with respect to the catalytic complex, the actin filament should swing around with it. Since the filament is very long compared to the ATP synthase (about 1 μm), its rotation should be visible in a fluorescence microscope. In other words, the fluorescently tagged actin filament, which is large enough to visualize in a light microscope, reports the rotation of the γ subunit.

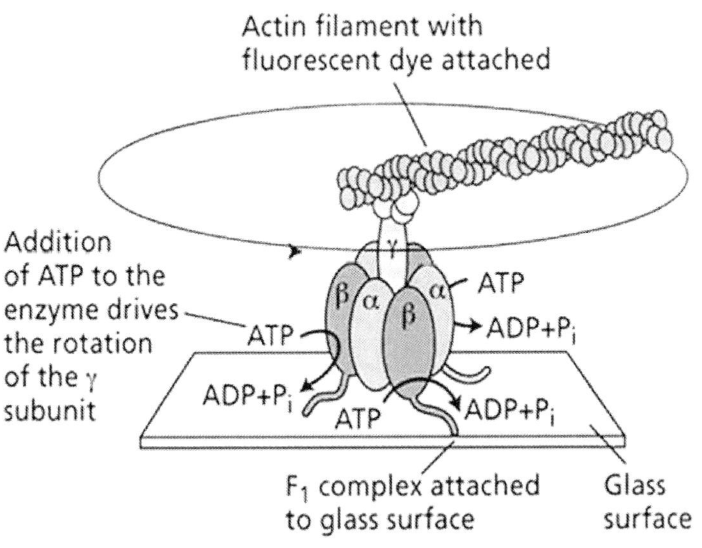

Figure : *A method for visualizing rotation of the g subunit. A fluorescently labelled actin filament was attached to one protruding end of the g subunit. The F_1 complex was then attached upside down to a coverslip. When ATP was added to the coverslip, the actin filament rotated. (After Noji et al. 1997.)*

The results were spectacular! When ATP was added to the modified enzyme, the actin filaments were seen to swing around in a circle at as much as 4 revolutions per second in a fluorescence microscope. To give some idea of scale, if you were a γ subunit, this rotation rate would be equivalent to swinging a several hundred meter-long rod around your head at 4 revolutions per second in water! However, the measured velocity is undoubtedly an underestimation of the actual velocity in vivo, because of the enormous torque required to swing such a large mass. Demonstration of the rotary motion of the γ subunit made it possible to put together a model of how the ATP synthase works.

For their contributions to elucidation of the mechanism of ATP synthesis, Paul Boyer and John Walker shared half the Nobel prize in physiology or medicine in 1997. The other half went to Jens Skou for his pioneering work on the K^+,Na^+-ATP synthase, the mammalian counterpart of the plant plasma membrane H^+-ATP synthase.

Transport Into and Out of Plant Mitochondria

Plant mitochondria must exchange metabolites and information (signal transduction) with the cytosol. To give some central examples, the end products of glycolysis, pyruvate and malate, need to be taken up, and citric acid cycle intermediates exported to be used as building blocks in biosynthetic processes.

Likewise, ADP and P_i must be taken up and ATP exported for the mitochondrion to fulfil its role as the powerhouse of the cell. To carry out the transport of this large number of molecules, plant mitochondria possess a family of mitochondrial carrier proteins.

These are transmembrane proteins with a common general structure and they typically transport relatively small charged molecules, like metabolites. In Arabidopsis there are almost 60 members, most of which are not functionally characterized, giving a hint to the complexity of the mitochondrial transport processes. However, mitochondria exchange more than metabolites, and therefore also other types of transporters are involved.

The outer membrane is permeable to molecules of less than 10 kDa, (i.e., all non-polymeric compounds). The permeability results from the presence of pores, which appear to be open in most *in vitro* experiments. Recent studies in animals have characterized pore-forming proteins in the mitochondrial outer membrane, called porins, which appear to regulate transport across the membrane.

Plant mitochondria also have porins but their characterization awaits further research. In our discussion here we will focus on the regulation of mitochondrial transport across the inner mitochondrial membrane. We will first consider the relatively few molecules that can move across the inner membrane without the help of a specific carrier protein. We will then discuss the various metabolite transporters and, finally, briefly look at the transport of proteins and RNA.

Small, Neutral, and Hydrophobic Molecules Diffuse across the Inner Membrane

Irrespective of size, molecules that either are charged or are very hydrophilic can only cross a lipid bilayer with the help of a specific protein carrier.

Small, *neutral* molecules can diffuse across the lipid bilayer and do not require protein transporters. Among those are O_2 and CO_2, two important molecules for respiration. O_2 is needed on the inner surface of the inner membrane where the active sites of both cytochrome *c* oxidase and the alternative oxidase are located.

The CO_2, produced as an end product in the citric acid cycle, needs to diffuse out of the mitochondrion, ultimately to be lost to the surroundings or re-fixed in the chloroplasts. In addition, however, it is possible that CO_2 can also be transported out of the mitochondrion as bicarbonate (HCO_3^-). Mitochondria are osmotically active and swell in a hypotonic medium. This means that the inner membrane is also permeable to H_2O, but it is not clear whether this occurs by diffusion or if aquaporins may be involved.

For technical reasons, it is of importance that several well-known inhibitors of mitochondrial function can diffuse into mitochondria because they are uncharged and hydrophobic. These are rotenone, the specific inhibitor of complex I, antimycin, the specific inhibitor of complex III, HCN, the protonated form of the cyanide (CN^-) inhibitor of complex IV (and other heme-containing enzymes), and oligomycin, the specific inhibitor of the F_oF_1-ATP synthase.

Transport of Inorganic Ions

Inorganic phosphate ($H_2PO_4^-$) is taken up into the matrix in exchange for a hydroxyl ion (HO^-). The energy for the uptake is supplied by the proton gradient (–pH). We will see below that the phosphate gradient, in turn, drives transport of other metabolites.

Several other inorganic ions—Mg^{2+}, Ca^{2+}, K^+, and NH_4^+—are involved in mitochondrial metabolism, but the mechanism by which they pass the inner membrane of plant mitochondria is not well understood. In mammalian mitochondria, uptake of K^+ through a K^+-specific channel, and extrusion via a K^+/H^+ antiporter, is involved in matrix volume regulation and cellular signalling (Garlid and Paucek 2003). A similar system of K^+ transport is present in plant mitochondria.

Both mammalian and plant mitochondria contain an uncoupling protein (UCP). This protein facilitates the movement of protons across the inner membrane and therefore uncouples electron transport and decreases ATP yield.

Transport of Adenylates

One of the important functions of mitochondria is to supply the cell with ATP. To do this, mitochondria have an adenine nucleotide transporter that facilitates the exchange of ADP^{3-} (inward) for ATP^{4-} (outward). This exchange is driven by the membrane potential (inside negative), since there is a net outward movement of one negative charge.

In combination with the phosphate carrier reaction, this means that the uptake of one ADP and one phosphate and the export of one ATP (i.e., the transport event connected to mitochondrial ATP synthesis) costs one vectorial proton—the electrical component for adenylate exchange and the "pH component for phosphate uptake.

Transport of Citric Acid Cycle Intermediates and Related Compounds

Pyruvate and malate, the end products of glycolysis, both have transporters in the inner membrane. Pyruvate is taken up in exchange for a hydroxyl ion, and malate^{2-} in exchange for $H_2PO_4^{2}$.

The transporter for malate also takes succinate^{2-} and is therefore called the dicarboxylate transporter. Malate^{2-} or succinate^{2-} can, in turn, exchange for citrate^{2-} on the tricarboxylate transporter. Malate^{2-} can also exchange for 2 oxoglutarate^{2-} via a special transporter. An interesting transporter, the oxaloacetate transporter, exchanges oxaloacetate^{2-} for malate^{2-} (or citrate^{2-}), which permits the shuttling of reducing equivalents.

The joint activity of these transporters allows plant mitochondria to import or export most citric acid intermediates. Finally, the glutamate/aspartate transporter facilitates the uptake of glutamate. In mammalian mitochondria, glutamate is taken up together with a proton, and it is thus driven by both the electrical and the chemical proton gradient. A similar transporter is present in plant mitochondria, but a proton co-transport has not been shown, so the transport may be passive in plants. Recently, several additional transporters for metabolites have been identified, and it is likely that certain setups of transporters are only expressed under special developmental conditions, like the mobilization of storage nutrients in developing seedlings.

In several cases the involvement of the mitochondrion in cellular metabolism implies that certain metabolites are taken up across the inner membrane, but the transporter has not yet been described. This is the case for the massive flux of glycine into—and CO_2, serine and NH_4^+ out of—mitochondria in C_3 leaves during photorespiration This is also the case for proline, which accumulates to high concentrations in the cytosol during episodes of plant water stress, and it is then broken down during the rehydration phase by a pathway involving two mitochondrial enzymes.

Transport of Coenzymes

Coenzymes are either synthesized inside the mitochondrion or imported. There is some evidence that, in addition to ATP, at least NAD^+, thiamine pyrophosphate, and coenzyme A are imported across the inner membrane.

The final step in ascorbic acid biosynthesis has recently been shown to be on the outer surface of the inner mitochondrial membrane. In view of the accumulating evidence that ascorbate has an important function in removing ROS inside the mitochondrion, we should expect to find an ascorbate transporter in the inner mitochondrial membrane.

The situation with folate (involved in 1C transfers) is the opposite to that with ascorbate: The only site of synthesis of many of the folate metabolites in the plant cell is the mitochondrial matrix. We can therefore expect that these molecules are exported via specific transporters.

Transport of Reducing Equivalents—Metabolic Shuttles

Mammalian mitochondria cannot oxidize cytosolic NADH or NAD(P)H directly, so they make use of "shuttle mechanisms." An exchange of metabolites by inner membrane transporters takes place through these shuttles so that a reduced compound moves into the matrix in exchange for a more oxidized compound moving out into the cytosol.

In this way, reducing equivalents are transported into the matrix ultimately to be oxidized by the respiratory chain. The shuttles can also be used to transport reducing equivalents in the opposite direction. Although plant mitochondria can oxidize cytosolic NADH and NAD(P)H directly by the two external NAD(P)H dehydrogenases, they can also use these two metabolic shuttles. The malate/oxaloacetate shuttle uses malate dehydrogenase in the cytosol and in the matrix to catalyze the interconversion of malate (reduced) and oxaloacetate (oxidized).

These two compounds are exchanged on the oxaloacetate transporter. This exchange, however, is not driven by the electrochemical proton gradient across the inner membrane, so it can only move reducing equivalents from a relatively reduced compartment to a relatively oxidized compartment. Since the mitochondrial matrix is more reduced than the cytosol, the oxaloacetate transporter is likely to work primarily in the export of reducing equivalents.

Under photorespiratory conditions, large amounts of glycine are oxidized in the mitochondria to give NADH while, at the same time, NADH is required in the peroxisomes to reduce hydroxypyruvate to glycerate. It is thought that as much as 50% of the matrix NADH is exported, probably via the malate/oxaloacetate shuttle.

The more complex malate/aspartate shuttle involves two matrix enzymes and two cytosolic enzymes as well as two inner membrane transporters. In mammals, it is powered by the electrochemical proton gradient in the direction of the import of reducing equivalents. However, since the glutamate/aspartate exchanger may be passive in plants, this shuttle in plants may similarly be independent of the electrochemical proton gradient. If so, it would (like the malate/oxaloacetate shuttle described above) only be able to function as an exporter of mitochondrial reductant.

Transport of Proteins—Protein Import

Plant mitochondria contain their own DNA (mtDNA) and ribosomes and are capable of synthesizing the proteins encoded in the mtDNA. However, plant mtDNA encodes only a few of the proteins found in the mitochondrion. The great majority of the mitochondrial proteins have to be imported from the cytosol where they are synthesized on free ribosomes. These are synthesized with a presequence or targeting peptide in the N-terminal end, which is recognized by receptors on the outer mitochondrial membrane.

The import of proteins requires a complex machinery with two multisubunit protein complexes—one in the outer membrane (TOM, transporter of the outer membrane) and one in the inner membrane (TIM, transporter of the inner membrane). Concomitant with, or immediately afterwards the import of the protein into the correct mitochondrial compartment, the targeting peptide is often cleaved off to yield the mature and active protein.

Transport of RNA

Mammalian mtDNA encodes the full complement of tRNA required for mitochondrial protein synthesis. In contrast, plant mtDNA only encodes some of the tRNAs required for protein synthesis within the mitochondrion. The missing tRNA must be imported by a yet-to-be-elucidated import mechanism.

The Genetic System in Plant Mitochondria has Some Special Features

As in plastids, the genetic system in mitochondria differs both structurally and functionally from the nuclear genome and its gene expression. On the other hand, mitochondria share several genetic features with bacteria. For example, several genes are transcribed together and the ribosomes are of a prokaryotic type. This is in line with the evolutionary origin of the mitochondrion as an endosymbiont of the a-subclass of Proteobacteria, a group including the intracellular pathogen *Ricketsia prowazekii*.

Plants are also distinctly different from other eukaryotes in their mitochondrial DNA (mtDNA) and its expression. For example, the plant mtDNA is very large and highly variable in size between species. The presence of several repeats enables the genome to recombine, changing the gene order or even causing a split into smaller subgenomic molecules, each carrying a subset of the genes. Though the size of mtDNA varies between plants, the number of genes appears to be similar in all plants.

In the fully sequenced 367 kilobasepair (kb) mtDNA of *Arabidopsis thaliana*, the known genes (constituting 65 kb) are scattered in a background of noncoding sequence. A similar situation can be found in the sequenced mtDNA of sugarbeet.

Plant mtDNA is quite different from the much smaller (16 kb) single mtDNA molecule of mammals in which every base pair has a function in coding for protein or mediating transcription and replication, and neither introns nor repeats exist. The expression of the mitochondrial genes in plants is carried out by RNA polymerases that are unlike the nuclear polymerases, but related to the enzymes in bacteriophages.

For several mitochondrial genes, these phage-type polymerases start up transcription from multiple promoters. Many genes in plant mtDNA are broken up into exons by introns. The introns are removed in the processing of the primary RNA into a mature mRNA that can be translated into the correct polypeptide.

The introns in mitochondria belong to the so-called self-splicing type II group, in which splicing is defined by the intron structure. The splicing can also work in the *trans* mode, in which it can join segments from separate RNA molecules, and thus assemble protein-coding reading frames even if the exons are distributed on different primary transcripts. This allows individual exons of a gene to reside independently in the genome, even in different subgenomic DNA molecules.

Because mitochondrial genes are often co-transcribed with unrelated genes into a polycistronic primary transcript, RNA processing also includes endonucleolytic reactions, so that each mature mRNA encodes only one protein. Like mRNAs for nuclear encoded genes, mitochondrial mRNAs are polyadenylated. However, instead of stabilizing (the case for nuclear encoded mRNAs), the mitochondrial mRNA polyadenylation mediates degradation. Instead, stem-loops stabilise the mRNA in plant mitochondria.

In plant mitochondria, an additional RNA processing event takes places—RNA editing. In this process, certain cytidines (C) are deaminated into uridines (U). Rarely, U is aminated to C instead.

This means that the coding sequence is changed at the RNA level, in order to restore the correct information for translation where the DNA has diverged. Editing sites mainly occur in DNA encoding protein sequences and functional RNA, and are rare in noncoding mtDNA. In principle, the higher the importance of an individual base, the higher the probability that it constitutes an editing site. Despite the considerable amount of bases that can be edited (up to 15% of the bases of a gene), the functional relevance of this process and the consequent divergence in mtDNA sequence is not understood. Mitochondrial RNA editing has been found in land plants, but not in algae. It also occurs, though to a lesser extent, in plastids.

The mitochondrial genome encodes only a small proportion of the mitochondrial proteins; less than 40 as compared to up to thousands of nuclear-encoded mitochondrial proteins. Through evolution, genes have been transferred from the mitochondrion to the nucleus. This is a process that still takes place, at least in plants.

For example, subunit 2 of cytochrome oxidase is encoded by the mitochondrial genome in all investigated eukaryotes except for some legumes and protists. Within the legume family, several intermediates of the gene transfer have been found. For example, a functional gene can be found in the mitochondrial genome and an unexpressed one in the nuclear genome, or vise versa.

In order for a mitochondrial gene to be successfully transferred to the nucleus, the protein-coding frame must be integrated into the chromosome, and several other modifications must be made.

For example: (1) a promoter for nuclear transcription must be inserted at the beginning of the gene, (2) the mRNA should acquire sequences specifying the addition of a poly-A tail, and (3) a targeting sequence is necessary for the cytosolically synthesized protein to be imported into the mitochondrion. For proper function, elements regulating the expression in different cell types and under different conditions must also be included.

Since the coding sequence for most mitochondrial genes in plants is edited at the RNA level, the mitochondrial genomic sequence will not code for a functional protein if inserted into the nucleus.

Therefore, the DNA that is integrated into the nuclear genome must in most cases be made by reverse transcription from mitochondrial mRNA. Recently, upon sequencing the nuclear genome of *Arabidopsis thaliana*, a copy of most of the mtDNA was found integrated in a nuclear chromosome, indicating that large segments of DNA can move between cellular compartments.

However, most of these mitochondrial genes transferred to the nuclear genome cannot produce functional proteins unless bases corresponding to editing sites are corrected by mutation.

However, this is unlikely to take place before other mutations have further disrupted the sequence. As opposed to other mitochondrial genomes, plant mitochondria utilize the universal genetic code. Thus, this factor will not prevent or delay the transfer of genes between plant organelles as it may do in other eukaryotes.

The special features of the mitochondrial genetic system in plants also have consequences for the study of mitochondrial genes. Due to RNA editing and *trans*-splicing, a reading frame is more difficult to identify based only on genomic sequence. Also, protein sequences in databases are automatically translated from genomic DNA sequences without compensating for RNA editing in many cases. This protocol would not work for mitochondrial genes.

Does Respiration Reduce Crop Yields?

Plant respiration can consume an appreciable amount of the carbon fixed each day during photosynthesis over and above the losses due to photorespiration. To what extent can changes in a plants respiratory metabolism affect crop yields?

Attempts to establish a quantitative relationship between respiratory energy metabolism and the various processes going on in the cell have led to a break-up of respiration into two components. In growth respiration, reduced carbon is processed to bring about the addition of new biomass. The other component, maintenance respiration, is needed to keep existing, mature cells in a viable state. Utilization of energy by maintenance respiration is not well understood, but estimates indicate that it can represent more than 50% of the total respiratory flux.

Although numerous questions remain regarding these issues, there are several empirical examples of relations between plant respiration rates and crop yield. In the forage crop, perennial ryegrass (*Lolium perenne*), yield increases of 10% to 20% were correlated with a 20% decrease in the leaf respiration rate.

Similar correlations have been found for other plants, including corn (*Zea mays*) and tall fescue (*Festaca arundinacea*). However, later investigations have shown that "selection for low rates of mature leaf respiration is not an appropriate method to select for high-yielding cultivars in perennial grasses".

Although there is a potential for increasing crop yield through reduction of respiration rates, a better understanding of the sites and mechanisms that control plant respiration is needed before such changes can be exploited commercially in a systematic fashion by plant physiologists, geneticists, and molecular biologists.

Furthermore, much remains to be learned about the general applicability of such observations and the conditions under which slower-respiring lines could be at a disadvantage, causing a reduction in crop yields rather than an increase.

Two recent studies illustrate the present difficulty of predicting the effect of directed changes at the molecular level on plant productivity:

Nunes-Nesi et al. (2005) found that the productivity of transgenic tomato plants with a reduced activity of mitochondrial malate dehydrogenase was increased as compared to the wild type. Although the respiratory activity of mitochondria isolated from the transgenic plants was unchanged or higher than in the wild type, the rate of leaf respiration was reduced in the transgenic plants. Photosynthesis was markedly increased in the transgenic lines, possibly linked to increased levels of ascorbate.

Sieger et al. (2005) studied the effect of the alternative oxidase on the growth of tobacco cell cultures. Considering the energy-wasteful nature of the alternative oxidase one might expect that a cell line lacking the enzyme would grow faster than the wild type. Secondly, alternative oxidase has a role in the response of plants to oxidative stress, so a cell line lacking the alternative oxidase would also be expected to handle abiotic stress less well.

The results were surprising—the transgenic cell line with a very low expression of alternative oxidase grew as fast as the wild-type cells under normal, nutrient-sufficient conditions, and faster than the wild type under conditions of macronutrient limitation (low phosphate or low nitrogen).

It appears that the alternative oxidase is an important factor in modulating the growth rate in response to nutrient availability. Plant respiration involves an intricate metabolic network of interacting pathways and a complex regulation of gene expression and enzymatic activities. We clearly need to know much more about these interactions before we can predict the effects of changing the expression of single genes on plant productivity.

4

Stomatal Conductance and Yields of Irrigated Crops

In the last few decades, the breeding of agricultural crops for higher yields has been very successful. Breeders usually select for high-yielding genotypes empirically, without particular attention to specific plant traits that might be conducive to higher yields.

However, comparing old, low-yielding lines of any crop with advanced, high-yielding lines shows clearly that many morphological, physiological, and biochemical traits have been altered by the intense selection pressures for higher yields.

These changes indicate that selection for high-yielding genotypes has generated indirect selection pressures on the altered traits.

If we exclude effects of genes that regulate the expression of two or more unrelated traits (pleiotropic genes), the study of high-yielding lines might reveal specific traits and genes associated with the higher yield. With this information, breeders could explicitly select for yield-enhancing traits to improve yields further.

Changes in plant traits associated with yield increases can be studied conveniently in a historical series of a crop. Historical series include successive commercial releases within a breeding program, with each member of the series representing an incremental increase in yield. In studies aimed at characterizing relationships between specific traits and higher yields, all the members of the series are grown together under the same conditions, and the traits of interest, such as plant architecture or photosynthetic rates, are measured in parallel with yields.

Recent studies of this type focusing on historical series of Pima cotton (*Gossypium barbadense*) and bread wheat (*Triticum aestivum*) have shown a remarkable positive correlation between yield increases and increases in stomatal conductance.

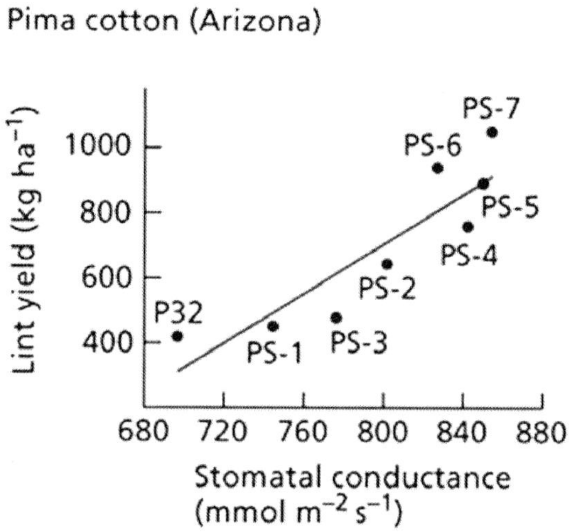

Figure: Stomatal conductance has increased in parallel with agronomic yields in irrigated Pima cotton (Gossypium barbadense) selected for higher yields at a high temperature. The figure shows the relationship between lint yield and stomatal conductance in a historical series of Pima cotton grown in Arizona. The abbreviations P32 and PS-1 through PS-7 designate successive commercial releases between 1949 and 1996.

Pima cotton is a high-quality fibre cotton grown in hot environments under intensive irrigation. The historical series used in the Pima studies included eight members, released between 1949 and 1996, and encompassing a nearly threefold increase in lint yield.

The yield increase attained in each commercial release was accompanied by a corresponding increase in stomatal conductance. Stomatal conductance increased by about 30 mmol m^{-2} s^{-1} for each 100 kg ha^{-1} increase in yield.

Genetic crosses between low-and high-yielding Pima strains have shown that stomatal conductance in Pima cotton has a clear-cut genetic component. The studies with stomata of Pima cotton are the first to show a clear-cut genetic regulation of stomatal conductance proper, of the stomatal response to temperature, and of proton pumping rates of isolated guard cells.

In the absence of explicit selection for higher stomatal conductance in the Pima breeding program, what could be the nature of the indirect selection pressures that caused the increases in conductance paralleling the increases in yield? Higher stomatal conductance increases CO_2 diffusion into the leaf and favours higher photosynthetic rates.

Higher photosynthetic rates could in turn favour a higher biomass and higher crop yields. Advanced Pima lines show a higher photosynthetic capacity than older, low-yielding lines, but photosynthetic rates measured in the same leaves used to measure stomatal conductance were not positively correlated with yields.

Thus, higher stomatal conductance appears to favour higher yields by a mechanism not directly related to photosynthesis.

Evapotranspiration at the leaf surface lowers leaf temperature, and higher stomatal conductance enhances this leaf cooling. Optimal daytime temperatures for growth, photosynthesis, and reproduction in Pima cotton are below 30°C, while afternoon air temperature in the hot Pima-growing areas often exceeds 40°C.

Evaporative cooling of the leaves thus reduces the gap between optimal growth temperatures and air temperature, and provides heat resistance. Studies with the Pima historical series showed that, because of leaf cooling, leaf temperatures were several degrees lower than air temperature, and leaf and canopy temperature were lower in the advanced, high-yielding lines than in low-yielding lines.

The same relationship among yields, stomatal conductance, and leaf temperature was found in a historical series of bread wheat grown in the warm Yaqui Valley of northwestern Mexico. These studies indicate that selection for higher yields in irrigated crops grown at high temperatures imposes indirect selection pressures for high stomatal conductance that lowers leaf temperature and appears to reduce deleterious effects of heat stress on critical flowering and fruiting stages, thus resulting in higher crop yields.

If selection for higher yields generates indirect selection pressures for higher stomatal conductance, can breeders explicitly select for higher conductance and thus obtain lines with higher yields? Studies with Pima cotton have shown that selection of high-conductance F_2 progeny from a cross between high-and low-conductance parents produces high-conductance F_4 lines with higher lint yields than low-conductance lines have. These experiments indicate that high stomatal conductance could be used as a selection trait for high yields in irrigated crops grown at high temperature.

The observed increases in stomatal conductance in crops grown at high temperature also have valuable implications for models developed to predict global climate changes, which are expected to occur because of increases in atmospheric CO_2. Under atmospheric CO_2 concentrations that are twice as high as the present ones, evapotranspiration would decrease over the continents and air temperatures would increase significantly over the tropical land masses, amplifying the changes resulting from atmospheric radiative effects. However, these models have yet to take into consideration the effect of higher temperatures on the stomatal control of evapotranspiration. Incorporation of the stomatal response to temperature in climate change models is likely to improve the accuracy of the predictions generated by the models. These studies illustrate both the complex interactions between physical and biological factors in the biosphere, and the advantage of a thorough understanding of the physiological properties of plants for efforts aimed at increasing agronomic yields and at improving models that predict global climate changes.

The GABA Shunt Pathway Integrates Stress Metabolic Responses

Not all enzymes participating in metabolic pathways have identical temperature responses, and differential thermostabilities of enzymes may also effect specific steps in metabolism before HSPs acting as molecular chaperones can restore activity. Thus, heat stress can lead to changes in metabolism that result in the accumulation of some metabolites and the reduction of others. Such changes can dramatically alter the function of metabolic pathways and lead to imbalances that are difficult to correct. This problem can be extremely critical when the levels of regulatory metabolites are forced outside of normal ranges so that regulatory homeostasis cannot function properly to restore normal levels.

Additionally, the consumption—and production—of protons during metabolic reactions and/or changes in the activity of membrane-associated proton-transporting ATPases and other membrane-associated proton transporters that move protons out of the cytosol either into the apoplast or vacuoles, can also cause a lowering of the pH of the cytosol, leading to additional metabolic perturbations during stress. Metabolic adaptation to the stress results in the amelioration of such stress effects on metabolism.

One of the metabolic adaptations that plants make to heat stress leads to the accumulation of the nonprotein amino acid, 4-aminobutyric acid (GABA). During episodes of heat stress, GABA accumulates to levels six-to tenfold higher than in unstressed plants. GABA is made from the amino acid, L-glutamate, in a single reaction catalyzed by the enzyme glutamate decarboxylase (GAD). GAD is one of several enzymes whose activity is modulated by the calcium regulatory protein, calmodulin. In the presence of calcium ions, calmodulin is converted into an active state that binds to GAD activating the enzyme, which leads to an increase in the rate at which GABA is produced. The precise mechanism leading to increases in cytosolic calcium have not been directly demonstrated, but studies using inhibitors of calcium ion channels implicate either the plasma membrane or tonoplast calcium channels. Alternatively, the efficiency of one or more of the CAX gene products (calcium proton antiporters) may be reduced as the proton gradient collapses during heat stress, leading to elevated levels of calcium in the cytosol. These increases, in conjunction with calmodulin, activate GAD, producing the high-temperature-induced synthesis of GABA.

GABA is metabolized inside the mitochondria by a GABA aminotransferase (GAT) that utilizes pyruvate and GABA as substrates generating alanine and succinate semialdehyde. This latter compound is then converted into succinate by succinate semialdehyde dehydrogenase (SSA-DH). The function of the mitochondrial components of this pathway leads to the reduction of pyruvate inside the mitochondria and the bypass of citric acid cycle metablism between pyruvate and succinate. This has the effect of bypassing three of the major NADH-producing reactions inside the mitochondria and constitutes a major metabolic adjustment to cellular heat stress that can dramatically alter the redox status of the cell.

Although GABA is an important signalling molcule in mammalian brain tissue, there is no evidence that it directly functions as a signalling molecule in plants. However, the synthesis of GABA consumes protons and thus the GAD-catalyzed reaction is one of several reactions involved in the regulation of cytosolic pH. The function of GAD aids the cell during the heat stress episode by maintaining cytosolic pH in a normal physiological range. Additionally, Arabidopsis mutants of SSA-DH demonstrate increased hydrogen peroxide production during episodes of heat stress and show increased programmed cell death. Thus, the function of the GABA metabolic pathway appears to modulate the oxidative status of the cell during heat stress, and may be involved in the production of a cell signal, effectively modulating PCD.

Membrane Lipids and Low Temperatures

The importance of membrane lipids for tolerance to low temperatures has been demonstrated by work with mutant and transgenic plants in which the activity of particular enzymes led to a specific change in membrane lipid composition independent of acclimation to low temperature.

In one study, the genes used to modify the lipid composition of thylakoid membranes encoded glycerol-3-phosphate acyltransferases (GPATs). These enzymes select fatty acids for incorporation into PG; GPATs from different plant species have different selectivity for saturated and unsaturated fatty acids, leading to the differences in composition.

The degree of fatty acid saturation of phosphatidylglycerol (PG) is particularly important to the chilling resistance of chloroplasts. In one experiment, tobacco was transformed with an acyltransferase gene from either squash (chill sensitive) or *Arabidopsis* (chill tolerant) to modify the fatty acid saturation of PG in the thylakoid membranes of the chloroplasts. In transgenic tobacco transformed with the squash gene, which increased the proportion of high-melting-point (trans-monounsaturated and saturated) molecular species of PG, chilling tolerance decreased (88% inhibition of photosynthesis following 4 hours at 1°C).

Conversely, when tobacco was transformed with the *Arabidopsis* chill-tolerant gene, the proportion of high-melting-point molecular species of PG decreased, and tolerance of chilling was greater than in wild-type plants.

Ice Formation in Higher-Plant Cells

During the initial phase of heat loss from tissue, when the temperature drops below the freezing point of the cytosol and the vacuole without ice forming, the liquid phase in the cytosol and vacuole is said to be supercooled. As the temperature drops further, ice forms in the intercellular spaces, and heat energy (0.33 kJ or 80 calories per gram) is released as a result of the latent heat of fusion of water. At this stage, the temperature of the tissue reflects the balance between heat gain from ice formation and heat loss to the environment. As a result, when ice first forms, the temperature rises rapidly (point B in the figure), and it remains at that level until all the extracellular water is frozen (point C in the figure). When that point is reached, heat release stops, and the temperature begins to fall again. The release of heat energy during ice formation (points B to C in the figure) is the basis for the common practice of spraying crops with water during frost: As long as the water continues to freeze extracellularly, it releases heat that prevents intracellular freezing.

The formation of ice within the intercellular spaces of cells that are sensitive to freezing is not lethal, but extended exposure to freezing temperatures causes water vapour to move through the plasma membrane from the unfrozen protoplast to the cell wall, causing ice crystals to grow within the intercellular spaces. This slow dehydration concentrates solutes within the protoplast, depressing the freezing point by 2 to 3°C. As the temperature continues to drop, a second phase of release of the heat of fusion of water is detectable (points from D to E on the figure). This phase reflects a series of small freezing events: Each "spike" of heat release is thought to represent the freezing of cell protoplasts and coincides with loss of viability. The formation of ice crystals on cell walls or in the protoplasm requires the presence of ice nucleation points on which crystals can be initiated and grow.

In some species, acclimation confers an ability to suppress ice nucleation in the protoplast, allowing deep supercooling to many degrees below the freezing point without ice formation. However, deep supercooling of intracellular water has a lower limit of about –40°C, the temperature at which ice forms spontaneously. (The homogeneous ice formation temperature of pure water droplets is –38.1°C, but the presence of solutes lowers this temperature.) At or below –40°C, ice crystals form without nucleation points, and intracellular freezing and cell death are unavoidable. Species that tolerate temperatures below –40°C under natural conditions do so not by supercooling, but by tolerating gradual dehydration.

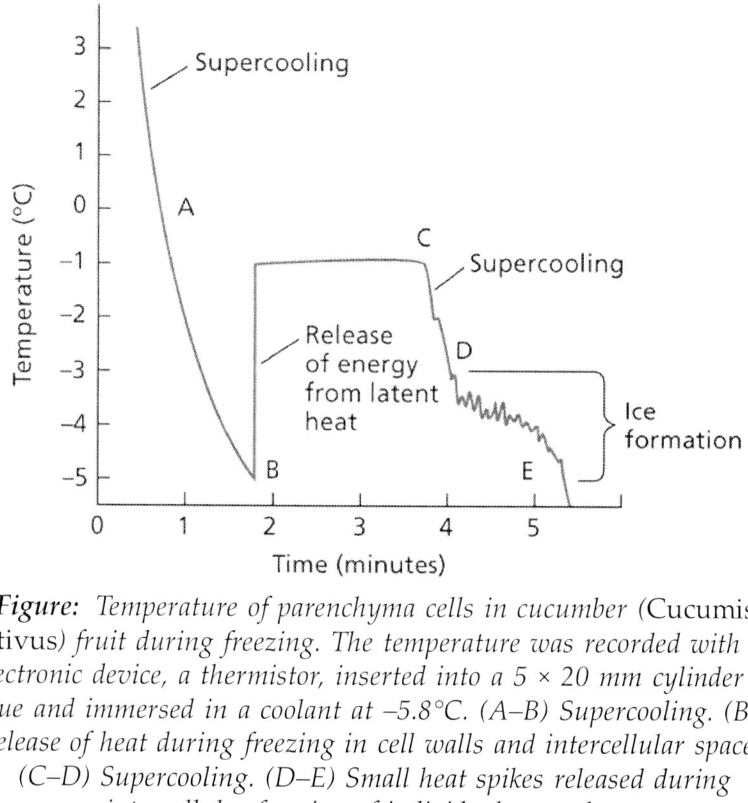

Figure: *Temperature of parenchyma cells in cucumber (*Cucumis sativus*) fruit during freezing. The temperature was recorded with an electronic device, a thermistor, inserted into a 5 × 20 mm cylinder of tissue and immersed in a coolant at –5.8°C. (A–B) Supercooling. (B–C) Release of heat during freezing in cell walls and intercellular spaces. (C–D) Supercooling. (D–E) Small heat spikes released during intracellular freezing of individual protoplasts.*

Plants produce compounds that protect cells against intracellular ice formation. Some of these compounds, such as sugars, amino acids, and other solutes, induce supercooling in the plant's tissues. Overwintering plants also produce antifreeze proteins and other low molecular weight compounds. These compounds provide freeze tolerance by inhibiting ice crystal growth and the nucleation of ice crystals.

Ca^{2+} Signalling and Activation of the Salt Overly Sensitive (SOS) Signal Pathway

Recent studies have identified components of the SOS signal pathway as genetic loci that mediate salt sensitivity. Three genetically linked loci (*SOS1, SOS2* and *SOS3*) constitute determinants of a stress-signalling pathway that controls ion homeostasis and salt tolerance.

- SOS3 encodes an EF-hand type Ca^{2+} binding protein with sequence similarity to the regulatory B subunit of calcineurin (protein phosphatase 2B) and neuronal Ca^{2+} sensors. SOS3 has three EF-hand structures, and the interaction of SOS3 with SOS2 kinase leading to activation of SOS2 activity is Ca^{2+} dependent. The function of SOS3 as a salt tolerance determinant *in planta* is dependent on N-myristoylation.
- SOS2 is a serine/threonine protein kinase that is similar to yeast SNF1 and mammalian AMPK protein kinases. Although all the outputs of the SOS signal pathway have yet to be completely defined, genetic and biochemical evidence indicates that the principal effectors regulated by the pathway are transport systems involved in ion homeostasis at least some of which are the same ones discussed above.
- The plasma membrane Na^+/H^+ antiporter, SOS1, is controlled both at the transcriptional and post transcriptional levels by the SOS signalling pathway. Functional disruption of *AtHKT1* was shown to suppress the salt-sensitive phenotype of *sos3*, indicating that the SOS pathway negatively controls this Na^+-influx system. Additionally, the SOS signalling pathway negatively controls expression of vacuolar AtNHX family members.

These studies illustrate the combined use of genetic, physiological, and biochemical approaches for the elucidation of molecular determinants of ion homeostasis and their signal-mediated regulation in plants. Functional genetic studies are investigating the role of these transport proteins in tolerance to salinity stress.

Na^+ Transport across the Plasma Membrane and Vacuolar Compartmentation

The energy-dependent transport of Na^+ from the cytosol of plant cells across the plasma membrane is mediated by a Na^+/H^+ antiporter. This antiporter is known as SOS1 and it is presumably representative of a family of SOS1 genes. *Arabidopsis sos1* mutants are much more sensitive to growth inhibition by salt than are wild type plants. Interestingly, SOS1 gene expression is not regulated by NaCl stress; but it rather depends on other components of the SOS signalling pathway.

Several of the transporters that mediate Na^+ uptake into the cytosol across the plasma membrane have now been identified. In patch clamp studies with isolated membranes, sodium has been shown to compete with K^+ in inward rectifying K^+ channels. Other reports have suggested that K^+ outward-rectifying channels might facilitate Na^+ influx as well. The high affinity K^+ transporter (HKT) from wheat and the low affinity cation transporter (LCT1) from wheat facilitate Na^+ transport in patch clamp studies with isolated membranes, providing evidence to support their possible function in Na^+ uptake. Patch clamp studies have shown the presence of Ca^{2+}-insensitive-sensitive Na^+ conductances in isolated plant plasma membranes. The Ca^{2+}-sensitive component of Na^+ uptake has been attributed to nonselective cation channels (Schachtman et al. 1997), while *in planta* studies using *Arabidopsis* mutants have directly demonstrated the function of AtHKT1 as a Ca^{2+}-insensitive Na^+-uptake system (Rus et al. 2001). These studies strongly implicate *HKT*1 in the direct control of at least some Na^+ influx. Since *HKT* transcripts are expressed predominantly in the roots, it most probably functions in the control of Na^+ flux into the xylem for export to the shoot.

Vacuolar compartmentalization of Na^+, energized by the ÄpH across the tonoplast membrane, also involves a family of Na^+/H^+ antiporters. *Arabidopsis AtNHX*1 is the prototype for such transporters. Transgenic *Arabidopsis* and tomato plants overexpressing *AtNHX*1 exhibit enhanced salt tolerance. The *Arabidopsis AtNHX*1 family include six loci showing amino acid sequence similarity and topological complementarity to *AtNHX*1, including 12 transmembrane spanning domains and the presence of amiloride binding sites. The *AtNHX*1 family of proteins plays a critical role in the vacuolar compartmentalization of Na^+, which minimizes cytotoxicity and facilitates osmotic adjustment.

Gene Transfer and Stress Tolerance

The transfer of genes suspected to mediate stress tolerance to a test plant where they are overexpressed or altered in some way has given scientists an important new tool with which to study genes involved in stress tolerance. Results of the first study in which transgenic plants were reported to show altered stress tolerance after transfer of a foreign gene appeared in 1993.

The transferred gene encoded the enzyme mannitol dehydrogenase, and its expression led to the accumulation of mannitol in the transgenic plants. Transfer of other genes generated transgenic plants that accumulated other solutes, such as trehalose, proline, and fructan.

Interestingly, the amount of accumulated solutes was insufficient to osmotically balance the applied osmotic stress, yet all of the transgenic plants exhibited some increased stress tolerance. These surprising results are not fully understood, but the newly acquired solutes might protect the transgenic plant against free radicals generated in response to stress, rather than function as osmotic-balancing solutes.

Improved resistance to drought, salinity, and extreme temperatures has been observed in transgenic plants that express/overexpress genes regulating osmolytes, specific proteins, antioxidants, ion homeostasis, transcription factors, and membrane composition.

Plant Physiology

Plant physiology is a subdiscipline of botany concerned with the function, or physiology, of plants. Closely related fields include plant morphology (structure of plants), plant ecology (interactions with the environment), phytochemistry (biochemistry of plants), cell biology, and molecular biology.

Fundamental processes such as photosynthesis, respiration, plant nutrition, plant hormone functions, tropisms, nastic movements, photoperiodism, photomorphogenesis, circadian rhythms, environmental stress physiology, seed germination, dormancy and stomata function and transpiration, both part of plant water relations, are studied by plant physiologists.

Scope

The field of plant physiology includes the study of all the internal activities of plants—those chemical and physical processes associated with life as they occur in plants. This includes study at many levels of scale of size and time. At the smallest scale are molecular interactions of photosynthesis and internal diffusion of water, minerals, and nutrients. At the largest scale are the processes of plant development, seasonality, dormancy, and reproductive control. Major subdisciplines of plant physiology include phytochemistry (the study of the biochemistry of plants) and phytopathology (the study of disease in plants). The scope of plant physiology as a discipline may be divided into several major areas of research.

First, the study of phytochemistry (plant chemistry) is included within the domain of plant physiology. In order to function and survive, plants produce a wide array of chemical compounds not found in other organisms. Photosynthesis requires a large array of pigments, enzymes, and other compounds to function. Because they cannot move, plants must also defend themselves chemically from herbivores, pathogens and competition from other plants.

They do this by producing toxins and foul-tasting or smelling chemicals. Other compounds defend plants against disease, permit survival during drought, and prepare plants for dormancy. While other compounds are used to attract pollinators or herbivores to spread ripe seeds.

Secondly, plant physiology includes the study of biological and chemical processes of individual plant cells. Plant cells have a number of features that distinguish them from cells of animals, and which lead to major differences in the way that plant life behaves and responds differently from animal life. For example, plant cells have a cell wall which restricts the shape of plant cells and thereby limits the flexibility and mobility of plants. Plant cells also contain chlorophyll, a chemical compound that interacts with light in a way that enables plants to manufacture their own nutrients rather than consuming other living things as animals do.

Thirdly, plant physiology deals with interactions between cells, tissues, and organs within a plant. Different cells and tissues are physically and chemically specialized to perform different functions. Roots and rhizoids function to anchor the plant and acquire minerals in the soil. Leaves function to catch light in order to manufacture nutrients. For both of these organs to remain living, the minerals acquired by the roots must be transported to the leaves and the nutrients manufactured in the leaves must be transported to the roots. Plants have developed a number of means by which this transport may occur, such as vascular tissue, and the functioning of the various modes of transport is studied by plant physiologists.

Fourthly, plant physiologists study the ways that plants control or regulate internal functions. Like animals, plants produce chemicals called hormones which are produced in one part of the plant to signal cells in another part of the plant to respond.

Many flowering plants bloom at the appropriate time because of light-sensitive compounds that respond to the length of the night, a phenomenon known as photoperiodism. The ripening of fruit and loss of leaves in the winter are controlled in part by the production of the gas ethylene by the plant.

Finally, plant physiology includes the study of how plants respond to conditions and variation in the environment, a field known as environmental physiology. Stress from water loss, changes in air chemistry, or crowding by other plants can lead to changes in the way a plant functions. These changes may be affected by genetic, chemical, and physical factors.

Biochemistry of Plants

The list of simple elements of which plants are primarily constructed—carbon, oxygen, hydrogen, calcium, phosphorus, etc.—is not different from similar lists for animals, fungi, or even bacteria. The fundamental atomic components of plants are the same as for all life; only the details of the way in which they are assembled differs.

Despite this underlying similarity, plants produce a vast array of chemical compounds with unusual properties which they use to cope with their environment. Pigments are used by plants to absorb or detect light, and are extracted by humans for use in dyes.

Other plant products may be used for the manufacture of commercially important rubber or biofuel. Perhaps the most celebrated compounds from plants are those with pharmacological activity, such as salicylic acid (aspirin), morphine, and digitalis. Drug companies spend billions of dollars each year researching plant compounds for potential medicinal benefits.

Constituent Elements

Plants require some nutrients, such as carbon and nitrogen, in large quantities to survive. Such nutrients are termed macronutrients, where the prefix *macro*-(large) refers to the quantity needed, not the size of the nutrient particles themselves. Other nutrients, called micronutrients, are required only in trace amounts for plants to remain healthy. Such micronutrients are usually absorbed as ions dissolved in water taken from the soil, though carnivorous plants acquire some of their micronutrients from captured prey.

The following tables list element nutrients essential to plants. Uses within plants are generalized.

Macronutrients. (Necessary in large quantities)

Element	*Form of uptake*	*Notes*
Nitrogen	NO_3^-, NH_4^+	Nucleic acids, proteins, hormones, etc.
Oxygen	O_2 H_2O	Cellulose, starch, other organic compounds
Carbon	CO_2	Cellulose, starch, other organic compounds
Hydrogen	H_2O	Cellulose, starch, other organic compounds
Potassium	K^+	Cofactor in protein synthesis, water balance, etc.
Calcium	Ca^{2+}	Membrane synthesis and stabilization
Magnesium	Mg^{2+}	Element essential for chlorophyll
Phosphorus	$H_2PO_4^-$	Nucleic acids, phospholipids, ATP
Sulfur	SO_4^{2-}	Constituent of proteins and coenzymes

Micronutrients. (Necessary in small quantities)

Element	*Form of uptake*	*Notes*
Chlorine	Cl^-	Photosystem II and stomata function
Iron	Fe^{2+}, Fe^{3+}	Chorophyll formation
Boron	HBO_3	Crosslinking pectin
Manganese	Mn^{2+}	Activity of some enzymes
Zinc	Zn^{2+}	Involved in the synthesis of enzymes and chlorophyll
Copper	Cu^+	Enzymes for lignin synthesis
Molybdenum	MoO_4^{2-}	Nitrogen fixation, reduction of nitrates
Nickel	Ni^{2+}	Enzymatic cofactor in the metabolism of nitrogen compounds

Pigments

Among the most important molecules for plant function are the pigments. Plant pigments include a variety of different kinds of molecules, including porphyrins, carotenoids, and anthocyanins. All biological pigments selectively absorb certain wavelengths of light while reflecting others. The light that is absorbed may be used by the plant to power chemical reactions, while the reflected wavelengths of light determine the colour the pigment will appear to the eye.

Chlorophyll is the primary pigment in plants; it is a porphyrin that absorbs red and blue wavelengths of light while reflecting green. It is the presence and relative abundance of chlorophyll that gives plants their green colour. All land plants and green algae possess two forms of this pigment: chlorophyll *a* and chlorophyll *b*. Kelps, diatoms, and other photosynthetic heterokonts contain chlorophyll *c* instead of *b*, while red algae possess only chlorophyll *a*. All chlorophylls serve as the primary means plants use to intercept light in order to fuel photosynthesis.

Carotenoids are red, orange, or yellow tetraterpenoids. They function as accessory pigments in plants, helping to fuel photosynthesis by gathering wavelengths of light not readily absorbed by chlorophyll. The most familiar carotenoids are carotene (an orange pigment found in carrots), lutein (a yellow pigment found in fruits and vegetables), and lycopene (the red pigment responsible for the colour of tomatoes). Carotenoids have been shown to act as antioxidants and to promote healthy eyesight in humans.

Anthocyanins (literally "flower blue") are water-soluble flavonoid pigments that appear red to blue, according to pH. They occur in all tissues of higher plants, providing colour in leaves, stems, roots, flowers, and fruits, though not always in sufficient quantities to be noticeable. Anthocyanins are most visible in the petals of flowers, where they may make up as much as 30% of the dry weight of the tissue. They are also responsible for the purple colour seen on the underside of tropical shade plants such as *Tradescantia zebrina*; in these plants, the anthocyanin catches light that has passed through the leaf and reflects it back towards regions bearing chlorophyll, in order to maximize the use of available light.

Betalains are red or yellow pigments. Like anthocyanins they are water-soluble, but unlike anthocyanins they are indole-derived compounds synthesized from tyrosine. This class of pigments is found only in the Caryophyllales (including cactus and amaranth), and never co-occur in plants with anthocyanins. Betalains are responsible for the deep red colour of beets, and are used commercially as food-colouring agents. Plant physiologists are uncertain of the function that betalains have in plants which possess them, but there is some preliminary evidence that they may have fungicidal properties.

Signals and Regulators

Plants produce hormones and other growth regulators which act to signal a physiological response in their tissues. They also produce compounds such as phytochrome that are sensitive to light and which serve to trigger growth or development in response to environmental signals.

Plant Hormones

Plant hormones, also known as plant growth regulators (PGRs) or phytohormones, are chemicals that regulate a plant's growth. According to a standard animal definition, hormones are signal molecules produced at specific locations, that occur in very low concentrations, and cause altered processes in target cells at other locations. Unlike animals, plants lack specific hormone-producing tissues or organs. Plant hormones are often not transported to other parts of the plant and production is not limited to specific locations.

Plant hormones are chemicals that in small amounts promote and influence the growth, development and differentiation of cells and tissues. Hormones are vital to plant growth; affecting processes in plants from flowering to seed development, dormancy, and germination. They regulate which tissues grow upwards and which grow downwards, leaf formation and stem growth, fruit development and ripening, as well as leaf abscission and even plant death.

The most important plant hormones are abscissic acid (ABA), auxins, gibberellins, and cytokinins, though there are many other substances that serve to regulate plant physiology.

Photomorphogenesis

While most people know that light is important for photosynthesis in plants, few realize that plant sensitivity to light plays a role in the control of plant structural development (morphogenesis). The use of light to control structural development is called photomorphogenesis, and is dependent upon the presence of specialized photoreceptors, which are chemical pigments capable of absorbing specific wavelengths of light.

Plants use four kinds of photoreceptors: phytochrome, cryptochrome, a UV-B photoreceptor, and protochlorophyllide *a*. The first two of these, phytochrome and cryptochrome, are photoreceptor proteins, complex molecular structures formed by joining a protein with a light-sensitive pigment. Cryptochrome is also known as the UV-A photoreceptor, because it absorbs ultraviolet light in the long wave "A" region. The UV-B receptor is one or more compounds that have yet to be identified with certainty, though some evidence suggests carotene or riboflavin as candidates. Protochlorophyllide *a*, as its name suggests, is a chemical precursor of chlorophyll. The most studied of the photoreceptors in plants is phytochrome. It is sensitive to light in the red and far-red region of the visible spectrum. Many flowering plants use it to regulate the time of flowering based on the length of day and night (photoperiodism) and to set circadian rhythms.

It also regulates other responses including the germination of seeds, elongation of seedlings, the size, shape and number of leaves, the synthesis of chlorophyll, and the straightening of the epicotyl or hypocotyl hook of dicot seedlings.

Photoperiodism

Many flowering plants use the pigment phytochrome to sense seasonal changes in day length, which they take as signals to flower. This sensitivity to day length is termed photoperiodism. Broadly speaking, flowering plants can be classified as long day plants, short day plants, or day neutral plants, depending on their particular response to changes in day length. Long day plants require a certain minimum length of daylight to initiate flowering, so these plants flower in the spring or summer. Conversely, short day plants will flower when the length of daylight falls below a certain critical level. Day neutral plants do not initiate flowering based on photoperiodism, though some may use temperature sensitivity (vernalization) instead.

Although a short day plant cannot flower during the long days of summer, it is not actually the period of light exposure that limits flowering. Rather, a short day plant requires a minimal length of uninterrupted darkness in each 24 hour period (a short daylength) before floral development can begin. It has been determined experimentally that a short day plant (long night) will not flower if a flash of phytochrome activating light is used on the plant during the night.

Plants make use of the phytochrome system to sense day length or photoperiod. This fact is utilized by florists and greenhouse gardeners to control and even induce flowering out of season, such as the *Poinsettia*.

Environmental Physiology

Paradoxically, the subdiscipline of environmental physiology is on the one hand a recent field of study in plant ecology and on the other hand one of the oldest. Environmental physiology is the preferred name of the subdiscipline among plant physiologists, but it goes by a number of other names in the applied sciences. It is roughly synonymous with ecophysiology, crop ecology, horticulture and agronomy. The particular name applied to the subdiscipline is specific to the viewpoint and goals of research. Whatever name is applied, it deals with the ways in which plants respond to their environment and so overlaps with the field of ecology.

Environmental physiologists examine plant response to physical factors such as radiation (including light and ultraviolet radiation), temperature, fire, and wind. Of particular importance are water relations (which can be measured with the Pressure bomb) and the stress of drought or inundation, exchange of gases with the atmosphere, as well as the cycling of nutrients such as nitrogen and carbon.

Environmental physiologists also examine plant response to biological factors. This includes not only negative interactions, such as competition, herbivory, disease and parasitism, but also positive interactions, such as mutualism and pollination.

Tropisms and Nastic Movements

Plants may respond both to directional and nondirectional stimuli. A response to a directional stimulus, such as gravity or sunlight, is called a tropism. A response to a nondirectional stimulus, such as temperature or humidity, is a nastic movement.

Tropisms in plants are the result of differential cell growth, in which the cells on one side of the plant elongate more than those on the other side, causing the part to bend toward the side with less growth. Among the common tropisms seen in plants is phototropism, the bending of the plant toward a source of light. Phototropism allows the plant to maximize light exposure in plants which require additional light for photosynthesis, or to minimize it in plants subjected to intense light and heat. Geotropism allows the roots of a plant to determine the direction of gravity and grow downwards. Tropisms generally result from an interaction between the environment and production of one or more plant hormones.

In contrast to tropisms, nastic movements result from changes in turgor pressure within plant tissues, and may occur rapidly. A familiar example is thigmonasty (response to touch) in the Venus fly trap, a carnivorous plant. The traps consist of modified leaf blades which bear sensitive trigger hairs. When the hairs are touched by an insect or other animal, the leaf folds shut. This mechanism allows the plant to trap and digest small insects for additional nutrients. Although the trap is rapidly shut by changes in internal cell pressures, the leaf must grow slowly in order to reset for a second opportunity to trap insects.

Plant Disease

Economically, one of the most important areas of research in environmental physiology is that of phytopathology, the study of diseases in plants and the manner in which plants resist or cope with infection. Plant are susceptible to the same kinds of disease organisms as animals, including viruses, bacteria, and fungi, as well as physical invasion by insects and roundworms. Because the biology of plants differs from animals, their symptoms and responses are quite different. In some cases, a plant can simply shed infected leaves or flowers to prevent to spread of disease, in a process called abscission. Most animals do not have this option as a means of controlling disease. Plant diseases organisms themselves also differ from those causing disease in animals because plants cannot usually spread infection through casual physical contact. Plant pathogens tend to spread via spores or are carried by animal vectors.

One of the most important advances in the control of plant disease was the discovery of Bordeaux mixture in the nineteenth century. The mixture is the first known fungicide and is a combination of copper sulfate and lime. Application of the mixture served to inhibit the growth of downy mildew that threatened to seriously damage the French wine industry.

5

Cellular Water Relations

Water is the most abundant constituent of all physiologically active plant cells. Leaves, for example, have water contents which lie mostly within a range of 55–85% of their fresh weight. Other relatively succulent parts of plants contain approximately the same proportion of water, and even such largely nonliving tissues as wood may be 30–60% water on a fresh-weight basis.

The smallest water contents in living parts of plants occur mostly in dormant structures, such as mature seeds and spores. The great bulk of the water in any plant constitutes a unit system. This water is not in a static condition. Rather it is part of a hydrodynamic system, which in terrestrial plants involves absorption of water from the soil, its translocation throughout the plant, and its loss to the environment, principally in the process known as transpiration.

The typical mature, vacuolate plant cell constitutes a tiny osmotic system, and this idea is central to any concept of cellular water dynamics. Although the cell walls of most living plant cells are quite freely permeable to water and solutes, the cytoplasmic layer that lines the cell wall is more permeable to some substances than to others.

If a plant cell in a flaccid condition—one in which the cell sap exerts no pressure against the encompassing cytoplasm and cell wall—is immersed in pure water, inward osmosis of water into the cell sap ensues. This gain of water results in the exertion of a turgor pressure against the protoplasm, which in turn is transmitted to the cell wall. This pressure also prevails throughout the mass of solution within the cell. If the cell wall is elastic, some expansion in the volume of the cell occurs as a result of this pressure, although in many kinds of cells this is relatively small.

If a turgid or partially turgid plant cell is immersed in a solution with a greater osmotic pressure than the cell sap, a gradual shrinkage in the volume of the cell ensues; the amount of shrinkage depends upon the kind of cell and its initial degree of turgidity. When the lower limit of cell wall elasticity is reached and there is continued loss of water from the cell sap, the protoplasmic layer begins to recede from the inner surface of the cell wall. Retreat of the protoplasm from the cell wall often continues until it has shrunk toward the centre of the cell, the space between the protoplasm and the cell wall becoming occupied by the bathing solution. This phenomenon is called plasmolysis.

In some kinds of plant cells movement of water occurs principally by the process of imbibition rather than osmosis. The swelling of dry seeds when immersed in water is a familiar example of this process.

Stomatal Mechanism

Various gases diffuse into and out of physiologically active plants. Those gases of greatest physiological significance are carbon dioxide, oxygen, and water vapour. The great bulk of the gaseous exchanges between a plant and its environment occurs through tiny pores in the epidermis that are called stomates. Although stomates occur on many aerial parts of plants, they are most characteristic of, and occur in greatest abundance in, leaves.

Transpiration Process

The term transpiration is used to designate the process whereby water vapour is lost from plants. Although basically an evaporation process, transpiration is complicated by other physical and physiological conditions prevailing in the plant. Whereas loss of water vapour can occur from any part of the plant which is exposed to the atmosphere, the great bulk of all transpiration occurs from the leaves. There are two kinds of foliar transpiration: (1) stomatal transpiration, in which water vapour loss occurs through the stomates, and (2) cuticular transpiration, which occurs directly from the outside surface of epidermal walls through the cuticle. In most species 90% or more of all foliar transpiration is of the stomatal type.

Transpiration is a necessary consequence of the relation of water to the anatomy of the plant, and especially to the anatomy of the leaves. Terrestrial green plants are dependent upon atmospheric carbon dioxide for their survival. In terrestrial vascular plants the principal carbon dioxide–absorbing surfaces are the moist mesophyll cells walls which bound the intercellular spaces in leaves. Ingress of carbon dioxide into these spaces occurs mostly by diffusion through open stomates. When the stomates are open, outward diffusion of water vapour unavoidably occurs, and such stomatal transpiration accounts for most of the water vapour loss from plants. Although transpiration is thus, in effect, an incidental phenomenon, it frequently has marked indirect effects on other physiological processes which occur in the plant because of its effects on the internal water relations of the plant.

Water Translocation

In terrestrial rooted plants practically all of the water which enters a plant is absorbed from the soil by the roots. The water thus absorbed is translocated to all parts of the plant. The mechanism of the "ascent of sap" (all translocated water contains at least traces of solutes) in plants, especially tall trees, was one of the first processes to excite the interest of plant physiologists.

The upward movement of water in plants occurs in the xylem, which, in the larger roots, trunks, and branches of trees and shrubs, is identical with the wood. In the trunks or larger branches of most kinds of trees, however, sap movement is restricted to a few of the outermost annual layers of wood.

Root pressure is generally considered to be one of the mechanisms of upward transport of water in plants. While it is undoubtedly true that root pressure does account for some upward movement of water in certain species of plants at some seasons, various considerations indicate that it can be only a secondary mechanism of water transport.

Upward translocation of water (actually a very dilute sap) is engendered by an increase in the negativity of water potential in the cells of apical organs of plants. Such increases in the negativity of water potentials occur most commonly in the mesophyll cells of leaves as a result of transpiration.

Water Absorption

The successively smaller branches of the root system of any plant terminate ultimately in the root tips, of which there may be thousands and often millions on a single plant. Most absorption of water occurs in the root tip regions, and especially in the root hair zone. Older portions of most roots become covered with cutinized or suberized layers through which only very limited quantities of water can pass.

Whenever the water potential in the peripheral root cells is less than that of the soil water, movement of water from the soil into the root cells occurs. There is some evidence that, under conditions of marked internal water stress, the tension generated in the xylem ducts will be propagated across the root to the peripheral cells. If this occurs, water potentials of greater negativity could develop in peripheral root cells than would otherwise be possible. The absorption mechanism would operate in fundamentally the same way whether or not the water in the root cells passed into a state of tension. The process just described, often called passive absorption, accounts for most of the absorption of water by terrestrial plants.

The phenomenon of root pressure represents another mechanism of the absorption of water. This mechanism is localized in the roots and is often called active absorption. Water absorption of this type only occurs when the rate of transpiration is low and the soil is relatively moist. Although the xylem sap is a relatively dilute solution, its osmotic pressure is usually great enough to engender a more negative water potential than usually exists in the soil water when the soil is relatively moist. A gradient of water potentials can thus be established, increasing in negativity across the epidermis, cortex, and other root tissues, along which the water can move laterally from the soil to the xylem.

Plant Mineral Nutrition

The relationship between plants and all chemical elements other than carbon, hydrogen, and oxygen in the environment. Plants obtain most of their mineral nutrients by extracting them from solution in the soil or the aquatic environment. Mineral nutrients are so called because most have been derived from the weathering of minerals of the Earth's crust. Nitrogen is exceptional in that little occurs in minerals: the primary source is gaseous nitrogen of the atmosphere.

Some of the mineral nutrients are essential for plant growth; others are toxic, and some absorbed by plants may play no role in metabolism. Many are also essential or toxic for the health and growth of animals using plants as food. Six basic facts have been established: (1) plants do not need any of the solid materials in the soil—they cannot even take them up; (2) plants do not need soil microorganisms; (3) plant roots must have a supply of oxygen; (4) all plants require at least 14 mineral nutrients; (5) all of the essential mineral nutrients may be supplied to plants as simple ions of inorganic salts in solution; and (6) all of the essential nutrients must be supplied in adequate but nontoxic quantities. These facts provide a conceptually simple definition of and test for an essential mineral nutrient. A mineral nutrient is regarded as essential if, in its absence, a plant cannot complete its life cycle.

Nutrients which plants require in relatively large amounts, that is, the essential macronutrients, are nitrogen, sulfur, phosphorus, calcium, potassium, and magnesium. Iron is not required in large amounts and hence is regarded as an essential micronutrient or trace element. With the progressive development of better techniques for purifying water and salts, the list of essential nutrients for all plants has expanded to include boron, manganese, zinc, copper, molybdenum, and chlorine. Evidence has accumulated in support of nickel being essential. In addition, sodium and silicon have been shown to be essential for some plants, beneficial to some, and possibly of no benefit to others. Cobalt has also been shown to be essential for the growth of legumes when relying upon atmospheric nitrogen. Claims that two other chemical elements (vanadium and selenium) may be essential micronutrients have still to be firmly established.

Mineral nutrients may be toxic to plants either because the specific nutrient interferes with plant metabolism or because its concentration in combination with others in solution is excessive and interferes with the plant's water relations. Other chemical elements in the environment may also be toxic. High concentrations of salts in soil solutions or aquatic environments may depress their water potential to such an extent that plants cannot obtain sufficient water to germinate or grow. Some desert plants growing in saline soils can accumulate salt concentrations of 20–50% dry weight in their leaves without damage, but salt concentrations of only 1–2% can damage the leaves of many species.

A number of elements interfere directly with other aspects of plant metabolism. Sodium is thought to become toxic when it reaches concentrations in the cytoplasm that depress enzyme activity or damage the structure of organelles, while the toxicity of selenium is probably due to its interference in metabolism of amino acids and proteins. The ions of the heavy metals, cobalt, nickel, chromium, manganese, copper, and zinc are particularly toxic in low concentrations, especially when the concentration of calcium in solution is low; increasing calcium increases the plant's tolerance. Aluminium is toxic only in acid soils. Boron may be toxic in soils over a wide pH range, and is a serious problem for sensitive crops in regions where irrigation waters contain excessive boron or where the soils contain unusually high levels of boron.

All plants grow poorly on very acid soils (pH d" 3.5); some plants may grow reasonably well on somewhat less acid soils. Several factors may be involved, and their interactions with plant species are complex. The harmful effects of soil acidity in some areas have been exacerbated by industrial emissions resulting in acid rain and in deposition of substances which increase the acidity on further reaction in the soil, with consequent damage to plants and animals in these ecosystems.

The elemental composition of plants is important to the health and productivity of animals which graze them. With the exception of boron, all elements which are essential for plant growth are also essential for herbivorous mammals. Animals also require sodium, iodine, and selenium and, in the case of ruminant herbivores, cobalt. As a result, animals may suffer deficiencies of any one of this latter group of elements when ingesting plants which are quite healthy but contain low concentrations of these elements. In addition, nutrients in forage may be rendered unavailable to animals through a variety of factors that prevent their absorption from the gut. Plants and animals differ also in their tolerance of high levels of nutrients, sometimes with deleterious results for grazing animals. For example, the toxicity of high concentrations of selenium in plants to animals grazing them, known as selenosis, was recognized when the puzzling and long-known "alkali disease" and "blind staggers" in grazing livestock in parts of the Great Plains of North America were shown to be symptoms of chronic and acute selenium toxicity.

Essential Points in the Nutrition of Plants

The Nature of Plant Growth Processes

The processes concerned in the growth of plants are the subjects of study by plant physiologists and plant biochemists. A comprehensive account of these processes is outside the scope of the present work, the special object of which is to deal with the outward and visible signs of imperfections in the plant's activities caused by faulty mineral nutrition. Nevertheless it is useful to have before us the general features of the main processes involved and to realize that the symptoms we shall be discussing later have a physiological basis, and are not direct and unchangeable signs of the specific deficiencies but result from the derangement of the complicated mechanism of the plant's vital activities.

The main processes involved in plant development may be summarized as follows:

Absorption: Intake of water and mineral elements by the root system.

Carbon assimilation or photosynthesis: Intake of carbon dioxide from the air by the leaves, and reaction of the gas with water in the leaf in the presence of the green chlorophyll to form sugar and free oxygen.

Formation of protoplasm: Protoplasm is the living material of the plant, consisting mainly of proteins, complex compounds of nitrogen built up by the plant from more simple compounds of this element.

Respiration: The combination of oxygen with various food substances synthesized by the plant, especially sugars, whereby energy is produced.

Transpiration: Loss of water from the plant, mainly from the leaves.

Translocation: The movement of materials within the plant.

Storage: Storage of reserve products in various organs and tissues.

During growth there are continuous processes of building up of complex compounds of carbon and nitrogen and breaking down of these into more simple substances, in which water and oxygen are intimately concerned. These processes together comprise plant metabolism.

In the course of the metabolic processes innumerable substances are formed, such as sugars, starch, cellulose, acids, lignin, tannins, amino acids, proteins, amides etc., and many plants also produce special products, as for instance nicotine in the tobacco plant.

For the normal functioning of the above processes there must be an adequate intake of water by the plant to maintain the plant cells in a more or less turgid condition and, since water is being continuously lost at a varying rate from the plant, intake and movement within the plant tissues must be capable of ready adjustment to these changes.

As a result of metabolic activities plants develop special organs of growth and reproduction, each of which has its special characters and makes particular demands on the nutrient supplies of the plant.

With all plants there are well defined seasonal growth cycles. Thus annuals, such as cereals, begin from the seed, give rise to seedlings, which later flower, form grain and ripen off, whilst perennial deciduous trees, such as apples, pears, etc., begin growth in the spring, using stored reserves of food, form leaves, make shoots, blossom and form fruits and subsequently shed their leaves, but meanwhile pass on reserve foods to various storage organs in preparation for the next season's growth. Coincident with these growth cycles there are well defined chemical cycles of nutrient elements and elaborated products in the leaves, stems and roots, etc. It will be shown later that these cycles are of great importance in considering p-deficiency effects and in diagnosing their causes.

The Plant Environment

Nutritional problems must be considered in relation to all the conditions in which plants live, and not merely in terms of the amounts of plant nutrients contained in or added to the soil. For example, those who are accustomed to growing plants know that the temperature must not be too low or no growth may result or that if too high the plants may be injured. An optimum temperature is usually recognized and this may vary according as to whether the plant is young or old. Similarly the importance of light is well known and plants may be put in special positions to obtain a maximum supply of light at one stage and may be shaded at another.

The actual duration of the daily period of illumination also affect growth and there are plants which are classified as requiring "long day" conditions to complete their growth cycles and others as needing "short day" conditions. If the special "long" or "short" day periods are not forthcoming for the respective classes of plants requiring these, their growth cycles are abnormal and they may fail entirely to produce flowers, grain or fruit.

The humidity of the atmosphere, as distinct from the water supply in the soil, is of importance in determining the water conditions within the plant, as these are dependent on both water intake by the roots and water loss from the leaves, the latter being largely influenced by the air humidity.

Even the presence of adequate quantities of plant nutrients in the soil is no guarantee that they will be absorbed by the plant roots. It will be shown later how these may be present in forms which are not available to the plants, but even when they would be considered as being present in suitable forms for absorption, other factors may prevent this taking place. An example of this latter condition is afforded in poorly aerated soils where lack of oxygen near the roots may prevent them from actively absorbing mineral nutrients

The problems of such influences in the plant environment as those just mentioned are complicated by the fact that they do not act independently, but their effects are modified by one another. Thus the effects of light intensity or period of daylight may vary with different temperature conditions.

The requirements of plants for different nutrients may be affected by conditions of light, temperature and water supply, and by other factors of the general environment. Thus the need for nitrogen may be less under conditions of relatively low light intensity whereas the need for potash in these circumstances may be greater, these facts being of importance in growing tomatoes under glass. The effect of nitrogen in relation to light may be shown by growing a plant under normal light conditions with insufficient nitrogen, when the leaves will show the well known symptoms of nitrogen deficiency-pale green, yellow, orange and red tints. If such a plant be then shaded, the leaves will turn a darker green and growth may be visibly increased. It can be shown that the lowered light conditions result in an increase of "soluble" a breaking down of proteins, thereby rendering the nitrogen of these available for growth processes.

This interrelationship of environmental factors is well illustrated by an experiment on apple trees at Long Ashton.

Bramley's Seedling trees were grown in compost in large pots and given a small dressing of a nitrogenous fertilizer. Some of the trees were grown in a specially constructed glass house and an equal number in an adjoining wire enclosure. The trees in the enclosure showed severe symptoms of nitrogen deficiency-pale green and yellow leaves, reddish brown barks and highly coloured, red fruits. The condition was corrected by further dressings of nitrogen. In contrast, the trees under glass, where the light was of less intensity and the temperature higher, made vigorous growth, carried large, green leaves and bore large, green fruits.

Iron and zinc deficiency symptoms may be less severe under conditions of low light intensity, whilst boron deficiency effects are less severe and magnesium deficiency effects are more pronounced in wet seasons than in dry ones.

The rate of water absorption is less at lower temperatures than at higher ones and efficient intake is also dependent on good aeration. These facts may result in a water deficit within plants growing in cold, wet soils when the air temperature is high. The raw materials needed for plant growth consist of carbon dioxide, which is obtained from the atmosphere through the stomata of the leaves, and water and the so-called mineral nutrients, which normally enter the plant through the medium of the roots.

The importance of water and carbon dioxide in the nutrition of plants will be apparent from the facts that water often comprises 80 to 90 % of the total weight of growing plants, and carbon and oxygen together may account for over 80% of their dry matter, i.e., the solid matter remaining after water is removed. As against these large amounts, the mineral nutrients, as measured by the ash content of the plants, i.e., the mineral residue obtained when the organic matter is destroyed by heat, often contribute from 5 to 15% of the dry matter.

It has been shown in recent years that certain organic compounds, known as "growth promoting substances" or "hormones", which occur in plants, and some of which are also present in soils and natural manures, are capable of producing marked growth responses, such as increased root growth, shoot and leaf curvatures, stimulation or suppression of buds, increased fruit setting, prevention of fruit abscission etc. They appear to perform important functions in the growth of plants. Examples of substances of this kind which can produce growth responses are 13 indole-acetic acid, 13 indole-butyric acid, phenyl acetic acid, A, naphthalene-acetamide, vitamin B_1

It is not at present clear to what extent growth substances are absorbed by plants from soils, although it has been shown that vitamin B_1, which occurs naturally in soils, can be obtained in this way.

The Mineral Nutrients

It has been shown in numerous researches that certain elements are necessary for the healthy growth of plants. They are sometimes spoken of as essential elements and, since some are needed in relatively large quantities and others in very small amounts, the former are referred to as "major" elements and the latter as "minor" or "trace" elements, or as micro nutrients.

The terms "major" and "minor" do not refer to the relative importance of the functions of the elements in plant growth, and for this reason the term "trace" element is preferable for the latter class. Major elements: Nitrogen, phosphorus, calcium, magnesium, potassium, sulfur.

Trace elements: Iron, manganese, boron, copper, zinc and molybdenum.

(Iron occupies an intermediate position and is usually included in the major elements group. In dealing with field problems it is more convenient to group it with the trace elements.)

In addition, there are, other elements, such as sodium, chlorine and silicon, which produce beneficial effects on the growth of certain plants but which have not so far been shown to be absolutely essential to growth. The element aluminium is of general occurrence in plants, but seems to be without direct nutritional value, although aluminium sulfate is used, because of its acidifying properties, to change the colour of hydrangeas growing on alkaline soils from pink to blue, and aluminium may also exert indirect influences on nutritional processes.

Other elements often occur in plants but they are not known to serve any useful function and frequently they act as plant poisons or toxins.

The nutrient elements can only be absorbed by plants when present in certain forms: nitrogen from nitrates and ammonium salts; phosphorus from phosphates; calcium, magnesium and potassium from their salts; sulfur from sulfates; iron from ferrous or ferric salts (more readily from ferrous salts); manganese from manganous salts; boron from borates; copper and zinc from their salts, and molybdenum from molybdates.

There may appear to be certain exceptions to this statement in practice. For instance, nitrogen may be applied to a soil as "organic" nitrogen, as in hoof meal or urea, and sulfur may be added as the element itself, as in flowers of sulfur, ground sulfur, etc. In such conditions the added materials are, however, converted into the nitrate and sulfate forms respectively by soil organisms before being absorbed by the plants.

Further points of importance in connection with the absorption of the mineral nutrients by plants are as follows:

(a) They must be absorbed from relatively dilute solutions or the plants will be! injured or even killed.

(b) Certain of the elements slow down the absorption of others into the plant, e.g., calcium slows down potassium and vice versa. The phenomenon is known as "antagonism".

(c) Healthy plants result when the nutrients are absorbed in certain relative proportions. When the proportions are suitable the nutrient medium is said to be "balanced". When ratios between nutrients are too wide, deficiency conditions are created. Thus if a high proportion of nitrogen to potassium is absorbed, the plant will suffer from potassium deficiency.

(d) Nutrients, even though present in the nutrient solution in satisfactory amounts and proportions, may not be absorbed by the plant unless the "reaction" of the solution as regards acidity and alkalinity is satisfactory. The reaction is measured in terms of the pH scale, which is merely a convenient notation for stating the conditions of acidity in the solution (strength or intensity of acidity, not total amount). The neutral point is represented by pH 7.0; below this value the solution is acid and above it is alkaline. Many crop plants prefer a reaction slightly on the acid side-pH 6.0 to 6.5 and extreme values are in the neighborhood of 4.0 on the acid side and 9.0 on the alkaline side.

(e) The nutrient medium must contain an adequate supply of oxygen, i.e., aeration must be satisfactory.

The Functions of The Mineral Nutrients

Knowledge of the main functions of the mineral nutrients is useful in helping us to understand the effects produced by deficiencies of any one of them.

A. Essential Major Elements

Nitrogen. Nitrogen is a major constituent of several of the most important substances, which occur in plants. It is of outstanding importance among the essential elements in that nitrogen compounds comprise from 40 to 50% of the dry matter of protoplasm, the living substance of plant cells. For this reason nitrogen is required in relatively large quantities in connection with all growth processes in plants. It follows directly from this that without an adequate supply of nitrogen appreciable growth cannot take place and that plants must remain stunted and relatively undeveloped when nitrogen is deficient.

Proteins, which are of great importance in many plant organs, e.g., seeds, are compounds of nitrogen whilst chlorophyll, the green colouring matter of the leaves, also contains the element.

From this latter fact it will be apparent that when nitrogen is deficient leaves will contain relatively little chlorophyll, and will thus tend to be pale green in colour.

In addition to the above substances, numerous other organic compounds of importance in plants, such as amino acids, amides, and alkaloids, are compounds of nitrogen.

Certain compounds of nitrogen are very mobile in plants, and this enables them readily to mobilize supplies of the element at vital growing points and to transfer stored supplies to points where they are most required. Such transference is common from old tissues to young growing points when supplies of the element are short. This mobility and re-utilization of nitrogen explains why deficiency symptoms of the element always appear first in the older parts of plants and why growing points are the last to be affected.

Phosphorus: This element, like nitrogen, is closely concerned with the vital growth processes in plants as it is a constituent of nucleic acid, and nuclei in which this occurs are essential parts of all living cells. Hence a deficiency of this element will also be expected to result in greatly restricted growth. Phosphorus is also of importance in seeds and in connection with the metabolism of fats. Compounds of phosphorus are concerned with the processes of respiration and with the efficient functioning and utilization of nitrogen. This relationship to nitrogen probably accounts for the fact that several of the symptoms of phosphorus deficiency are identical or similar to those which result from a deficiency of nitrogen, Phosphorus is also of special importance in the processes concerned in root development and the ripening of seeds and fruits.

Calcium: Calcium occurs in plants chiefly in the leaves and the amounts present in seeds and fruits are relatively low. One of its main functions is as a constituent of the cell wall, the middle lamella of which consists largely of calcium pectate. This function appears to be of fundamental importance since, if calcium is replaced by any other of the essential elements, such as magnesium or potassium, the organic materials and mineral salts in the cells are readily leached through the walls.

Other functions attributed to calcium are as follows:

It provides a base for the neutralization of organic acids; it is concerned with activities of growing points (meristems), especially with root tips; it may be of importance in nitrogen absorption.

Although a large proportion of the calcium contained in the plant may be soluble in water-as much as 60% in cabbage calcium does not appear to move freely from the older to the younger parts of plants, and hence young tissues contain lower proportions of calcium than older ones. This may explain why calcium deficiency effects begin at the tips of shoots.

Magnesium: The outstanding fact about magnesium is that it is a constituent of chlorophyll, and is essential to the formation of this pigment. As a result, when magnesium is deficient, one of the symptoms commonly shown by plants is chlorosis. Magnesium is also regarded as a carrier of phosphorus in the plant, particularly in connection with the formation of seeds of high oil content, which contain the compound lecithin.

The element seems to be very mobile within the plant, and when deficient is apparently transferred from older to younger tissues where it can be re-utilized in the growth processes. This agrees with the observation that signs of magnesium deficiency invariably make their appearance first on the oldest leaves and progress systematically from them towards the youngest ones.

Potassium: Unlike all the other major elements, potassium does not enter into the composition of any of the important plant constituents, such as proteins, chlorophyll, fats and carbohydrates, concerned in plant metabolism. For this reason its role is more difficult to determine, and in spite of much study it cannot be said that the functions of potassium are clearly understood.

The element is present in all parts of plants in large or fairly large proportions. It seems to be of special importance in leaves and at growing points, as these are especially rich in potassium. Probably the whole of the potassium in plants is present in soluble form, and most of it seems to be contained in the cell sap and cytoplasm.

It is outstanding among the nutrient elements for its mobility and solubility within the plant tissues, and these properties no doubt account for the ready way in which potassium can be re-utilized by young tissues when the element is in short supply.

Among the functions which have been attributed to potassium and the processes with which it may be concerned, the following may be mentioned: The formation of carbohydrates and proteins; the regulation of water conditions within the plant cell and of water loss by transpiration; as a catalyst and condensing agent of complex substances; as an accelerator of enzyme action; as contributing to photosynthesis through its radioactive properties.

It has been shown in many instances that the potassium content of plants is frequently much higher than is necessary for healthy growth, and it is generally considered that luxury absorption of potassium often takes place.

The great mobility of potassium in plants, its special importance for and its reutilization by young tissues, and its apparent functions as a regulator of plant processes on a large scale are in harmony with the observations that, when potassium is moderately deficient, the effects are seen first in the older tissues and progress from these towards the growing points, but, when the deficiency is acute, growing points are severely affected, and die-back and general collapse of the plants commonly occur.

Sulfur: Sulfur occurs in plants as a constituent of proteins, and of certain volatile compounds such as mustard oil. It seems to be connected with chlorophyll formation, although it is not a constituent of this substance. Its functions in connection with proteins and chlorophyll doubtless account for the similarity of its deficiency effects to those due to deficiency of nitrogen.

Essential Trace Elements

Iron: Iron is closely concerned with chlorophyll formation but is not a constituent of it. Its role appears in this connection to be that of a catalyst. As a result of this function of iron, chlorosis is invariably an outstanding symptom when the element is deficient. Iron may also act as a catalyst, in the role of an oxygen carrier, in respiration.

A point of great importance in connection with iron is its relative immobility in plant tissues. Its mobility seems to be affected by several factors, such as the presence of manganese, potassium deficiency and high light intensity. There is evidence that the amount of chlorophyll is related to "active" iron in plants.

It will thus be seen that so-called iron deficiency in the plant may in fact usually mean iron immobility. Lack of mobility may also account for the fact that iron deficiency is first shown in the younger tissues.

Manganese: The functions of manganese are regarded as being closely associated with those of iron and as being concerned with chlorophyll formation. Hence, when manganese is deficient, chlorosis is a common symptom. Manganese may decrease the solubility of iron by oxidation and hence, an abundance of manganese within the plant may lead to iron deficiency and chlorosis.

Manganese is regarded as having the functions of a catalyst; its activities being specially concerned with oxidation and reduction reactions within the plant tissues.

Boron: The exact role of boron is not known, but again the evidence as for the other trace elements, suggests that, its functions are those of a catalyst or reaction regulator. Particular effects attributed to boron are as follows:

It can delay the onset of calcium deficiency effects but cannot replace calcium; it tends to keep calcium soluble; it may act as a regulator of potassium/calcium ratios, and of the absorption of nitrogen; it may be concerned with the oxidation-reduction equilibrium in cells.

Such functions as the above accord with the results which follow from a deficiency of the element, when growth processes show sudden collapse and drastic derangements of metabolism occur.

Zinc and Copper: Although specific functions have not been determined for these elements, here again the evidence points to their roles as catalysts and regulators. Deficiencies of both are associated with chlorosis and a serious general collapse of vital growth processes.

Since catalysts are not used up in the chemical reactions which they promote, we can understand how it comes about that quite small or even minute quantities of the "trace elements", iron, manganese, boron, zinc and copper, may nevertheless be essential to the plant's health and growth.

Elements Sometimes Beneficial

Sodium: As sodium is not strictly an essential element it cannot be expected to have a specific role in the metabolic activities of plants.

Where sodium produces significant effects it is often regarded as a conserver of potassium and as being able partly to replace that element in its role. In no instance, however, has it been shown that sodium can wholly replace potassium where the latter is acutely deficient.

In such circumstances, sodium is ineffective as a substitute for potassium, even for sodium-loving plants, such as sugar beet, mangold and barley. Sodium seems to affect the water relations of plants and often enables sugar beet and other crops to withstand drought conditions which would otherwise produce severe adverse effects.

Chlorine: The evidence of the role of chlorine in plants is somewhat contradictory, and no general statement can be made. In tobacco it has been shown to increase the water content of the tissues and to affect carbohydrate metabolism, leading to an accumulation of starch in the leaves. The element is present in plants as chloride, and is wholly soluble.

Toxic Effects of Mineral Elements

Toxic effects in plants may be produced both by essential nutrient elements and non-essential elements.

In the first class, the major nutrients are much less toxic than the trace elements. Indeed, for the major nutrients, there exists a fair safety margin for excess or "luxury" consumption, but for the trace elements the margin is very narrow. Similar conditions exist in relation to non-essential elements thus some plants will tolerate fairly large amounts of elements such as sodium or chlorine, but are injured by relatively small amounts of elements like arsenic or chromium.

Two types of injury may occur: (1) An excess of one element may lead to a deficiency of another which ultimately results in a deranged metabolism, e.g., excess nitrogen or excess phosphorus may result in insufficient potassium and excess potassium may lead to deficiency of magnesium or calcium. This type of injury applies particularly to essential nutrients. (2) The presence of an element may directly injure the protoplasm and bring about the speedy death of the plant.

Signs of Faulty Nutrition in Plants

When plants are grown in unsuitable environments, including conditions of faulty mineral nutrition, they react to the particular defects in more or less specific ways. Thus, if light is insufficient, the green colouring matter of the leaf will be lacking and the leaves may be almost white (chlorotic), and the plant may be very spindly and "drawn" in appearance; if the temperature is too high, the growth may be rank and soft; if water is insufficient, growth may be restricted, the tissues woody and the green of the leaves show a bluish tint. Again, deficiencies and excesses of the individual elements produce characteristic effects on various organs of plants: foliage characters, including colour, density, size and shape of leaves; stem characters, such as thickness, colour and length of inter-nodes; root characters, such as colour, amount of fibre, abnormal thickening; blossom characters, including amount and time of opening of the flowers; fruit characters, such as size, colour, hardness and flavour.

Ability to recognize these particular effects forms the basis of the visual method of diagnosing plant deficiencies. Many of them can be readily learned and applied by practical farmers. Indeed, for many years progressive fruit growers in this country have used the leaf symptoms of deficiencies of nitrogen and potassium, and more lately of magnesium, as the main guide to manuring their fruit trees and bushes with these elements.

Detailed descriptions of deficiency symptoms and of the methods of using them in the field for diagnosing the manurial needs of crops.

Soils in Relation to the Supply of Mineral Nutrients

Since plants obtain their supplies of mineral nutrients from the soil, it is necessary to have some knowledge of soil conditions in order to understand the main problems of the supply of nutrient elements.

The Constitution of Soils

One of the first points which is brought out in any examination of a number of soils is their great diversity. In practical terms, soils may be good or bad; light or heavy; wet or dry; fertile or hungry; deep or shallow; black, red or brown; peaty, sandy, silty or clay; all terms denoting points of importance in relation to fundamental characters and to their practical utilization.

Soils are very complex bodies but, although showing such great variation, they all consist of five main components, mineral matter, organic matter, soil water, the soil atmosphere and a population of microorganisms.

Mineral Matter: The mineral matter furnishes the skeleton of the soil. It consists of materials which range in size from rock fragments and large pebbles to minute particles of clay which can be suspended in water for considerable periods. The mineral portion, which largely determines the texture of the soil, can be separated into its component parts or "fractions" by a combination of sieving and sedimentation in water. The British standard method of grading is as follows:

FRACTION	DIAMETER LIMITS
Stones	> 2.0 m.m.
Coarse Sand	2.0—0.2 m.m.
Fine Sand	0.2—0.02 m.m.
Silt	0.02—0.0002 m.m.
Clay	< 0.002 m.m.

A preponderance of the coarser fractions tends to make soils light, free working, freely drained and hungry, whilst where the finer fractions predominate (especially clay) the soils are heavy, sticky, difficult to work and retentive of water and manures.

The clay fraction is of special importance as it possesses colloidal properties, which give to soils many of their characteristic properties, such as swelling and shrinkage, holding up of water and absorption of mineral nutrients from manures. The clay colloid, together with the organic colloid (humus), acts as the soil storehouse for available plant nutrients.

The mineral matter of soils provides the main natural source of mineral nutrients, which become available to plants through the weathering of rock minerals and the complex processes of soil formation.

Organic Matter: The content of organic matter in soils shows great variation. Organic matter may comprise almost the whole of the solid matter of peats, while in mineral soils it may account for only 1 or 2% of the soil. It is usual in mineral soils in this country for the organic matter to range from 3 to 10% in the soil, the lower values occurring in arable soils and the higher ones in old gardens and under grass covers in pastures and meadows.

Plant residues, either naturally or artificially added, provide the source of organic matter in soils. These residues vary greatly in character and composition, depending on the plants from which they are derived, whether from trees, grasses, clovers, cereals, root and vegetable crops, etc., and on the nutrient supplying powers of the soils on which they are grown. In addition to plant residues, soil organisms and animal remains contribute to the organic matter. In the soil the fresh material undergoes chemical change, especially by the action of soil organisms, and when this has proceeded to the stage where the original cellular structure is no longer recognizable we speak of the brownish product as humus. The organic matter in the soil at any time thus consists of fresh and partly decomposed residues and humus.

The main points in the formation of humus are as follows:

The raw residues contain a variety of compounds, such as proteins and other nitrogenous materials, and sugars, starch, cellulose, tannins, lignin and nutrient salts. The soil organisms use compounds like sugars (carbohydrates) as sources of energy, ultimately oxidizing them to carbon dioxide. To do this they require a supply of nitrogen which they way take either from the nitrogenous compounds in the residues, or if this supply is insufficient they may use other nitrogen present in the soil. The most resistant to their attack of the compounds contained in the residues appears to be lignin, and it seems probable that the end-point of the more rapid decomposition processes humus is a complex mixture in which lignin derivatives preponderate.

During these processes the ratio of carbon to nitrogen (C/N) in the residues alters from about 40/1 to 10/1 so that the process entails a large reduction in carbon.

It will thus be seen that the organic matter of the soil supports the population of soil organisms, and of this population the bacteria alone may be of the order of 20 to 40 millions per gram of soil.

These organisms play a very great part in determining the availability of mineral nutrients in soils, by breaking down plant residues and also by providing carbon dioxide which, in combination with water, is of great importance for the weathering of the soil minerals. Certain of the soil organisms, especially bacteria, are also able to fix free nitrogen from the air and thus enrich the soil with this element. Other bacteria are present which cause the soil to lose nitrogen, and there are multitudes of protozoa which feed on the bacteria, and if too abundant render the soil unhealthy. A properly balanced soil population is requisite for good crops and disturbances of this balance either in a beneficial or deleterious direction may be brought about by soil treatment.

It is important to realize, however, that bacteria and other soil organisms do not work for the special benefit of crops, and that the changes they bring about are to meet their own particular needs, which are not always in harmony with the requirements of the crops. This point can be illustrated by digging into soil fresh straw or sugar and growing a crop on it immediately, when the crop will most probably show symptoms of acute nitrogen deficiency due to the fact that, in order to consume the large amount of carbohydrate contained in the straw or sugar, the organisms deplete the soil of its readily available nitrogen.

Organic matter produces both chemical and physical effects on soils. Like clay, the humified material possesses colloidal properties, depending on the minute size of the constituent particles and the enormous surface they collectively present. It can swell and shrink and absorb nutrient salts. Together with clay, humus coats the mineral particles and can bind these together forming "crumbs", thus giving "structure" to soils. Humus acts as a binding agent for the particles of coarse sandy soils and exerts an "opening" effect on close-textured clay soils. The binding properties are not associated with stickiness to the same extent as those of clay. Humus is superior to clay colloid as a storehouse of nutrient bases.

Soil Water: The solid particles of the soil, consisting of the mineral particles and the organic matter, being of varying dimensions, have spaces of varying sizes between them which are occupied by the soil water and the soil atmosphere. The total volume of these spaces in any soil is known as the pore space. It varies for different soils and even in the same soil under influences such as rainfall, drought, frost and management, but usually occupies from 30 to 60% of the soil volume. It will be clear that the amounts of water and air in the pore space are complementary and that when the soil is wet there will be correspondingly little air present and vice versa. A high water content in the soil thus means a condition of relatively poor aeration.

The manner in which water is held in soils and the way in which it moves through the soil mass provide difficult scientific problems. There is general agreement that in a wet soil a certain proportion of the water can be readily removed, and that some is held very strongly by forces in the soil. Moreover, it can be shown that only a proportion of the water in soils is available to plants. The movement of water through soils is also limited, and water tables at considerable depths below the surface cannot be relied upon to supply water to the surface layers of the soil by the well known capillary action which occurs in narrow tubes. Similarly, if a dry soil is wetted from the surface, the whole of the soil to a considerable depth does not become wetted through uniformly, but the top layers only are wetted, practically to full capacity, and the layers below remain dry.

This strictly limited movement of water through soils is of great practical importance, since it means that for practical purposes water does not move appreciably to the plant roots but the roots must go after the water.

The soil water contains the soluble products of the soil, and is the main nutrient medium for the plant roots. As such it is commonly called the soil solution. There is evidence that plants may also feed directly on nutrients contained in the soil colloids.

The soil solution is very dilute. Measurements made on displaced solutions show concentrations of soluble materials between 0.1 and 1.0%, whilst if drainage waters are analysed the corresponding values are of the order of 0.02 to 0.5%. In both, the proportions of calcium, nitrate, sulfate, carbon dioxide, silica and organic matter are relatively high, and there may be large proportions of chloride, sodium and magnesium, but potassium is always low and phosphate and ammonia, are present only in traces.

The soil organisms play an important part in providing soluble products for the soil solution. They liberate into the solution nitrate, sulfate and carbon dioxide and these, acting as acids, dissolve corresponding or equivalent amounts of base-forming elements, particularly calcium. Further changes then occur, the calcium in particular bringing into solution other "bases" such as magnesium and potassium by "base" exchange reactions with the colloidal clay and humus materials.

Under normal conditions, the whole of the nitrate and chloride is present in the soil solution, and they pass into the drainage water if not quickly absorbed by plants. Sodium is also mainly present in soluble form. Calcium, magnesium and potassium are held in considerable amounts in the soil colloids as "exchangeable bases", and in this form are less easily washed out of the soil.

Phosphorus is mostly present as insoluble compounds, as calcium phosphates in neutral and alkaline soils, and as iron phosphate in acid soils. Both potassium and phosphorus are only present in the soil solution in very small amounts, but apparently supplies of these elements in the solution are quickly replenished when removed by plants.

The amounts of iron, aluminium and manganese dissolved are largely dependent on the soil reaction (pH) and in strongly acid soils they may assume toxic concentrations to plants.

The conditions governing boron solubility are not known though, under drought conditions and at high pH values, availability to plants is reduced.

Copper and zinc (perhaps also manganese) appear to present complicated problems in which organic matter and the soil organisms are concerned.

Molybdenum may be more readily soluble in soils under alkaline than acid conditions, as it has been shown to be more readily available to pasture plants under the former conditions.

The Soil Atmosphere: The soil atmosphere or soil air, near the surface, has a composition similar to that of the ordinary air and diffusion between the two is rapid. Except near the surface the soil air is saturated with moisture.

The carbon dioxide content may fluctuate considerably, being increased by organic matter, by cropping, by high temperatures, which speed up the activities of soil organisms, and by the respiration of plant roots.

In poorly drained soils the concentration of carbon dioxide may be very high, with a correspondingly low oxygen content, and in extreme condition this may actually lead to inefficient absorption and even to root injury due to lack of oxygen. The fact that carbon dioxide accumulates and oxygen supply is depleted under conditions of poor drainage is important in considering deficiency problems in wet soils where practical results often appear contrary to expectation.

6

The Fluence Response of Phototropism

Phototropism involves two fundamental processes: blue-light perception and differential cell elongation. The identity of the blue-light photoreceptor involved in phototropism. There is now compelling evidence that two flavoproteins associated with the plasma membrane (PHOT1 and PHOT2) are the photoreceptors for phototropism. These proteins become phosphorylated in response to blue light. The differential cell elongation in response to unilateral light is thought to be caused by the lateral redistribution of auxin to the shaded side.

If phototropism were a simple photochemical reaction, one would expect the fluence response curve to rise from a threshold level of light to a plateau at which the photoreceptor becomes saturated. However, this is not what happens. The figure shows a diagram of a typical fluence response curve, which is similar to what has been observed in a wide range of species, including *Avena* coleoptiles and *Arabidopsis* stems.

Starting at the light threshold, there is a bell-shaped curve that has been termed the first positive curvature. The first positive curvature is followed by a neutral zone in which little or no measurable response is observed. Under certain conditions, *Avena* coleoptiles may actually bend away from the light in this range, a phenomenon referred to as the first negative curvature. Both the first positive and the first negative curvatures (neutral zone) are restricted to the coleoptile tip. At higher light fluences the curve ascends again, and this second bending response is termed the second positive curvature. The second positive curvature differs from the first positive curvature in that the bending occurs at the base rather than at the tip of the coleoptile.

Plant physiologists have long puzzled over the meaning of this complex relationship between light fluence and phototropism. Most studies of phototropism have been carried out in the range of the first positive curvature. The first positive curvature and the neutral zone obey the Bunsen–Roscoe reciprocity law, which states that the response will be the same for a brief exposure to bright light as for a longer exposure to dim light, as long as the product of time and the photon fluence rate are the same. Thus the first positive curvature is a response to short light pulses.

In contrast to the first positive curvature, the second positive curvature is a response to irradiations of long duration. The second positive curvature does not obey the reciprocity law; instead, it is proportional to the duration and the photon fluence rate. For the second positive response to occur, both the time threshold and the fluence threshold must be exceeded. Since these are the conditions that prevail normally in nature, phototropism in nature usually represents the second positive curvature.

What causes the gradient in auxin to form at the three fluences? It has long been assumed that the light gradient that is established across the coleoptile during unilateral irradiation with blue light must generate a biochemical gradient between the cells of the shaded and the irradiated sides, but the nature of this biochemical gradient has, until recently, eluded detection. With the discovery of blue light–induced phosphorylation of the PHOT1 and PHOT2 plasma membrane proteins, much work has focused on these photoreceptors.

Recently, Michael Salomon and his colleagues at the University of Munich demonstrated that unilateral blue light causes a lateral gradient in the phosphorylation of PHOT1 from the irradiated to the shaded side. Moreover, the concentration of PHOT1 declines exponentially from the tip of the coleoptile to the base. Studies on the rates of phosphorylation of the 116 kDa protein PHOT1 at the tip versus at the base of oat coleoptiles have led to a new model for the fluence response curve of phototropism.

According to this new model proposed by Salomon and colleagues, the gradient in concentration of NPH1 from tip to base makes the tip more sensitive to light than the base is. Thus, at threshold fluences, a gradient in phosphorylation occurs only at the tip, leading to a small bending response. Increasing the fluence increases the gradient of phosphorylation at the tip, until phosphorylation on the irradiated side saturates and a maximum phosphorylation gradient is obtained. This maximum gradient represents the peak of the first positive curvature.

As the fluence is increased beyond this point, phosphorylation increases on the shaded side due to the penetration of light, reducing the gradient. When phosphorylation of NPH1 is equal on both sides of the tip, no bending occurs (the neutral zone).

However, with continued exposure to light of sufficient fluence, phosphorylation increases at the base on the irradiated side. Phosphorylation at the base is dependent on both fluence and time, unlike phosphorylation at the tip which is strictly dependent on fluence. Thus, phosphorylation at the base appears to involve some type of time-dependent biochemical process. Since recent immunological studies have shown that the total amount of the 116 kD protein remains constant during phototropism, the time-dependent increase in phosphorylation at the base does not involve synthesis of new PHOT1 protein. Instead, a second photoreceptor (PHOT2), initially implicated in chloroplast movement in response to light, appears to function as an additional photoreceptor in second positive curvature conditions.

The model proposed by Salomon and colleagues requires further testing, and other interpretations are certainly possible. Nevertheless, the new model provides the first plausible explanation for the fluence response curve of phototropism. However, even if this model is confirmed, the question remains as to how a gradient in phosphorylation can bring about a gradient in auxin.

Gravity Perception without Statoliths in Chara

An alternative mechanism of gravity perception that does not involve statoliths has been documented in the giant-celled fresh water alga *Chara*. Despite the absence of any discernible sedimenting statoliths, *Chara* filaments show negative gravitropism. To explain this phenomenon, it has been suggested that the entire protoplast acts like a statolith, and that gravity is perceived at the interface between plasma membrane and cell wall.

According to the Plasmalemma Central Control (PCC) model, stretch-activated calcium channels are clustered around attachment centres connecting the cytoskeleton and the cell wall. These calcium channels are thought to open in response to tensions at the membrane caused by gravity-induced changes in the distribution of forces exerted by the protoplast, the cytoskeleton, or the cell wall. Something like the PCC model may explain how *Chara* and the starchless mutants of *Arabidopsis* are able to perceive and respond to gravity.

Cytokinins: Regulators of Cell Division

Cultured Cells Can Acquire the Ability to Synthesize Cytokinins

When cultured normal callus tissues of many species are subcultured repeatedly over a long period, they may become hormone autonomous; that is, they will grow on culture medium lacking an auxin or a cytokinin. This phenomenon is known as habituation.

There are several possible explanations for the habituation of plant tissues.

First, the tissues may acquire the capacity to synthesize their own cytokinin, and in some cases habituated callus tissues have been shown to contain cytokinins. Alternative explanations are that the rate of cytokinin destruction is reduced, or that the tissues become more sensitive to cytokinins, such that lower levels of the hormone are effective.

Frederick Meins and his colleagues in Basel, Switzerland, have shown that different tissues of tobacco plants exhibit different capacities to synthesize cytokinins in tissue culture. Tobacco pith tissue is cytokinin dependent but capable of being habituated. In contrast, tissues cultured from the cortex of the stem are cytokinin autonomous; they can be cultured indefinitely on medium containing auxin as the only hormone supplement. Finally, tobacco leaf tissue has an absolute requirement for cytokinins and cannot be habituated.

These observations indicate that the ability to synthesize cytokinins can be switched on or off during development, and that the on and off states can be stably maintained during cell divisions. Such developmental control of cytokinin synthesis may be important in morphogenesis.

Various Methods Are Used to Detect and Identify Cytokinins

Naturally occurring molecules with cytokinin activity may be detected and identified by a combination of physical methods and bioassays. The introduction of immunoaffinity purification methods combined with liquid chromatography and mass spectrometry, along with the use of isotopically labelled cytokinins as internal standards, has resulted in major advances in the measurement of endogenous cytokinins.

Historically, bioassay was the only method available for the identification of molecules that act as cytokinins. Cell proliferation bioassays using tobacco pith tissue or carrot taproot continue to be important because the ability of these tissues to initiate and sustain cell proliferation is proportional to the concentration of active cytokinin in an extract, provided the extract does not contain growth inhibitors.

In addition, some continuously cultured callus tissues of tobacco and soybean that cannot grow in culture without cytokinin have been used for cytokinin bioassays. All of these cytokinin-requiring tissues exhibit a linear increase in growth with increasing cytokinin concentration over a fairly broad range, although high concentrations of cytokinin usually inhibit growth.

Immunological methods are very useful for cytokinin identification and quantification. Researchers can produce antibodies against cytokinins by injecting rabbits or mice with cytokinin ribosides conjugated to a protein. Monoclonal antibodies also have been generated that are highly specific for individual cytokinins. These antibodies can be used to quantitate the amount of a cytokinin in a sample by means of a radioimmunoassay.

For cytokinin isolation, plant extracts are first fractionated, usually by high-performance liquid chromatography (HPLC), and the cytokinins in the fractions are detected and measured by means of a cytokinin radioimmunoassay, similar to the auxin radioimmunoassay.

The cytokinin antibodies can also be used to isolate the hormone from extracts by immunoaffinity chromatography. Immunological methods hold great promise for the identification and quantification of naturally occurring cytokinins because the antibodies are highly specific and more sensitive than most bioassays. Furthermore, these immunological methods are very rapid.

Cytokinins Are Also Present in Some tRNAs in Animal and Plant Cells

Most organisms contain a variety of modified bases in their tRNA. The normal RNA bases are incorporated into the newly synthesized tRNA chain, and then specific enzyme modify a small subset of the bases. As with the free cytokinins, Ä^2-isopentenyl groups are transferred to the adenine molecules. The enzyme catalyzing this reaction is called tRNA-IPT, which is weakly similar in its sequence to the plant cytokinin biosynthetic enzyme. The similarity was used to help identify the cytokinin biosynthetic enzyme from plants. The tRNA-IPT must be able to recognize a specific base sequence in the tRNA and transfer the isopentenyl group to the adenosine nearest the 3' end of the anticodon. It does not utilize free AMP as a substrate.

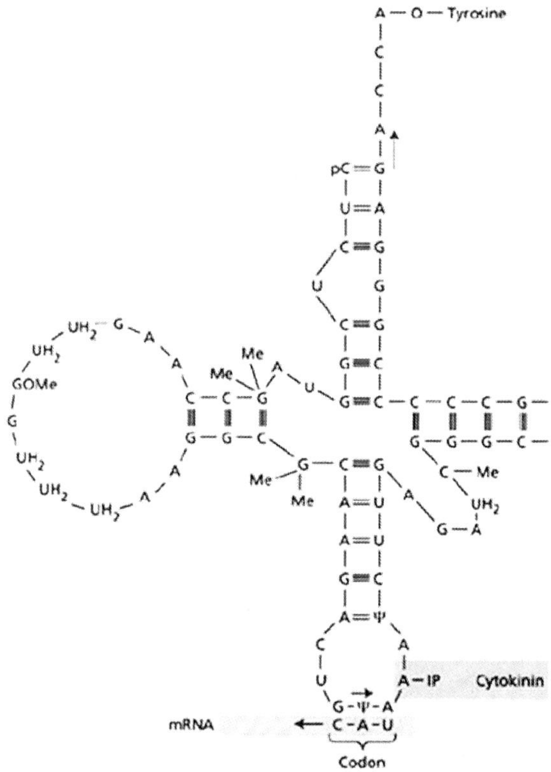

Yeast tRNA for tyrosine

Figure: Some tRNAs contain a cytokinin. The cytokinin in the examples shown is]^{9}R] iP. Although cytokinins are present in tRNA, tRNA is probably not the source of free cytokinins in plants.

The possibility that the free cytokinins are derived from tRNA has been explored extensively. Although the tRNA-bound cytokinins can act as hormonal signals for plant cells if the tRNA is degraded and fed back to the cells, it is unlikely that any significant amount of the free hormonal cytokinin in plants is derived from the turnover of tRNA.

The Ti Plasmid and Plant Genetic Engineering

The Ti plasmid (tumour-inducing plasmid) of *Agrobacterium tumefaciens* has been developed as a vehicle for introducing foreign genes into plants. When *Agrobacterium* infects plants, a region of the Ti plasmid called the T-DNA is taken up by the plant cell and incorporated into one of its chromosomes.

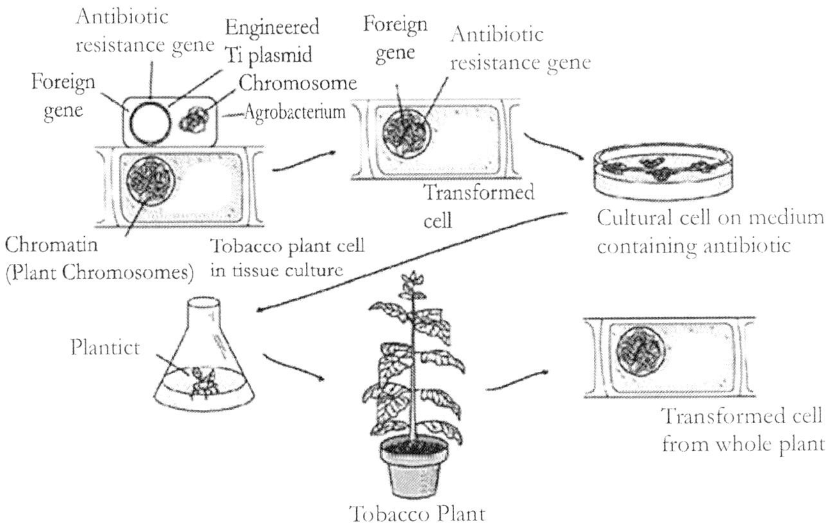

The genes in the T-DNA are referred to as phyto-oncogenes because they induce neoplastic, or tumour-producing, growth (Matzke and Chilton 1981). To use the Ti plasmid as a vector for introducing new genes into plants, it is necessary to disarm the plasmid so that it does not cause tumours.

Researchers accomplished this task by deleting the genes in the T-DNA that encode the enzymes controlling auxin and cytokinin synthesis. In addition, it is necessary to introduce a gene into the T-DNA that will enable the investigator to select the transformed cells. Genes for antibiotic resistance are normally used for this purpose.

A cloned gene can then be inserted into the T-DNA of the engineered Ti plasmid, and the plasmid can be used to infect cultured cells, leaf discs, or root slices. The infected cells are placed on a culture medium that contains auxin and cytokinin (to induce growth) and the antibiotic. Only the transformed cells can grow in the presence of the antibiotic, because they have received the T-DNA containing not only the foreign gene but also the gene for antibiotic resistance.

To obtain a plant containing the foreign gene, it is necessary to regenerate whole plants from the cultured, transformed cells. Fortunately, methods for accomplishing this regeneration have been developed for many plants, although not for all important crop species yet. The regeneration involves adjusting the ratio of cytokinin to auxin to stimulate both shoot and root formation.

Although the transformation of cereal crops with *Agrobacterium* and their regeneration from cultured cells has been difficult, some remarkable successes have resulted from this approach. Investigators have introduced numerous foreign genes into plants such as tobacco, including soybean storage protein genes. Genes for such desirable characteristics as disease resistance, herbicide resistance, and salt tolerance have been transferred to crop plants by these techniques, and commercial crops now are being grown with these genetically engineered strains.

Arabidopsis can also be transformed directly by exposing whole plants to a solution containing *Agrobacterium* that is carrying engineered or wild-type Ti plasmids. The plants must be treated in such a way to allow the *Agrobacterium* to enter tissue, either by applying a vacuum or by treating with detergents. The *Agrobacterium* penetrates the floral tissue and transforms the developing ovules. Isolation of seeds from these *Agrobacterium*-exposed plants yields up to 2% of the seeds that are transformed with the T-DNA. This approach is very useful for molecular genetic studies, such as for characterizing DNA sequences involved in the control of gene expression, or constructing large libraries of insertional mutants.

Cytokinin Can Promote Light-Mediated Development

The *det* mutants of *Arabidopsis* develop many of the characteristics of light-grown seedlings when they are germinated in the dark. The *det* mutants have short, unhooked hypocotyls and expanded cotyledons, and the apical meristem initiates leaves that subsequently expand without light. Although the *det* mutants do not develop full photosynthetic competence in the dark, there is partial development of the chloroplasts, including the expression of genes encoding many photosynthetic enzymes that are not present in the chloroplasts of dark-grown wild-type seedlings.

When wild-type *Arabidopsis* seedlings are germinated in darkness in the presence of cytokinin, they develop many of the characteristics of the *det* mutants: Hypocotyls shorten, cotyledons expand, the apical meristem initiates leaves, and there is partial development of the chloroplasts, including the synthesis of some photosynthetic enzymes. This effect is proportional to the cytokinin dosage given to the seedlings. Cytokinin is said to produce a *phenocopy* of the *det* mutations.

The ability of exogenous cytokinin to cause de-etiolation of dark-grown seedlings is mimicked by certain mutations that lead to cytokinin overproduction. Dark-grown *Arabidopsis* seedlings carrying the *amp1* mutation have zeatin levels that are approximately 2.5 times higher than those of the wild type and develop the characteristics of *det* mutant seedlings. Although the *det* mutants do not have elevated levels of cytokinins, they may have altered sensitivity to cytokinins, since *Arabidopsis* tissues from *det1* plants, when cultured in vitro, do not require cytokinin for growth and become green.

Although the functions of the genes altered by the *det* mutants are not completely understood, the *det2* mutation is in a gene required for the synthesis of brassinosteroids. Brassinosteroids and cytokinins may have antagonistic effects in regulating light-induced development. These results suggest that cytokinins may be part of the light signal transduction pathway that is responsible for the initiation of normal vegetative development and photosynthetic competence. However, the role for cytokinins in the developmental pathway initiated by light is not clearly established.

Ethylene: The Gaseous Hormone

Cloning of the Gene That Encodes ACC Synthase

A major hurdle in the understanding of ethylene biosynthesis was the difficulty in purifying the low abundance, unstable protein and the cloning of the corresponding gene. Researchers overcame this obstacle by raising antisera against partly purified ACC synthase from zucchini fruit slices that were induced to produce large amounts of ethylene. The resulting antisera contained a mixture of antibodies, including antibodies to ACC. The investigators then removed the contaminating antibodies by allowing them to bind to proteins extracted from zucchini tissue that lacked the enzyme. The key to the success of this procedure was the availability of tissue largely lacking the enzyme with which to subtract the non-ACS antibodies from the polyclonal antibody mixture. This circumvented the need for highly purified ACC synthase protein. The purified ACC synthase antibodies were then used to isolate the gene encoding the enzyme.

Once the gene encoding ACS was isolated, it could be used to express the ACC synthase protein in the bacterium *E. coli* at high levels and isolate purified ACC synthase in large amounts. The cloned zucchini gene also allowed genes encoding ACC synthase to be isolated from numerous other plant species, using low stringency hybridization and degenerate PCR. The deduced amino acid sequences of ACC synthase genes were found to show a high degree of similarity to the *aminotransferase* superfamily of enzymes. These enzymes catalyze the transfer of amino groups among amino acids and require pyridoxal phosphate as a cofactor.

Demonstration of the Dimeric Nature of ACC Synthase

The deduced crystal structure of both apple and tomato ACS reveals a homodimeric protein with shared active sites, which is similar to aminotransferases. The dimeric nature of ACC synthase has been confirmed by an elegant study involving functional complementation. In this study, researchers examined potential dimerization of ACS enzymes by expressing in *E. coli* two ACS proteins that harbour distinct active-site mutations, neither of which has activity on its own. If ACS acts as a monomer, then expressing the two different mutant versions should not result in an active enzyme, because each is inactive. Conversely, if the two inactive ACS subunits come together as a dimer, the different amino acid changes can complement each other, forming a functional enzyme. The researchers indeed found that functional ACS enzyme was formed when various pairs of mutants were expressed, which provides compelling evidence that ACS enzymes are able to act as both homo-and heterodimers with shared active sites.

Cloning of the Gene That Encodes ACC Oxidase

The gene that encodes ACC oxidase was cloned by a circuitous route. Researchers isolated a set of genes that are highly induced during tomato fruit ripening. They then blocked the expression of these genes by introducing an antisense version of each into transgenic tomatoes. Blocking of one such gene, *pTOM13*, repressed fruit ripening and inhibited ethylene production. Expression of *pTOM13* in *Xenopus* oocytes as well as in yeast conferred the ability to convert ACC to ethylene, indicating that the gene encodes an ACC oxidase enzyme. Using this gene, researchers then isolated ACC oxidase clones from numerous plant species.

ACC Synthase Gene Expression and Biotechnology

As we have seen, ACC synthase is a cytosolic enzyme that catalyzes the first committed step in ethylene biosynthesis in higher plants. It is a key regulatory enzyme that catalyzes the production of the ethylene precursor ACC from AdoMet (*S*-adenosylmethionine). However, this enzyme is difficult to purify because it is labile and present in low abundance in plant tissues. The molecular cloning and functional expression of ACC synthase in *E. coli* and yeasts have facilitated biochemical and structural studies of this enzyme.

ACC synthase cDNAs and genomic sequences have been cloned from numerous plant species. The emerging picture from the study of these genes is that ACC synthase is encoded by members of a divergent multigene family in which each gene is differentially regulated by various environmental and developmental factors during the growth of a plant. *Arabidopsis*, for example, has twelve and tomato has at least nine gene members, which are differentially regulated by inducers such as auxin, fruit ripening, and wounding.

The protein synthesis inhibitor cycloheximide is a potent inducer of many ACC synthase genes. Cycloheximide inducibility is a characteristic of *primary response genes* such as the early auxin-inducible genes. Two mechanisms have been proposed to explain this inducibility.

- First, the transcription of these genes may be under the control of a shortlived repressor protein whose synthesis is prevented by the addition of cycloheximide;
- Second, the mRNA transcripts of these genes may be shortlived and are stabilized when cycloheximide prevents the synthesis of a nuclease.

The induction of at least one of the genes for ACC synthase (*ACS2*) in *Arabidopsis* by cycloheximide appears to be due to transcriptional activation, and thus is likely regulated by a shortlived transcriptional repressor.

A clue to the signal transduction mechanism involved in the regulation of ACC synthase gene expression has come from experiments with lithium (Li^+). Li^+ is known to interfere with the phosphoinositide signalling pathway by inhibiting the activity of inositolphosphate phosphatase. Li^+ is also one of the strongest inducers of ACC synthase activity in plants. This finding suggests that the regulation of at least some ACC synthase genes may involve Ca^{2+} mobilization mediated by inositol trisphosphate (IP_3).

Because ethylene plays such an important role in fruit ripening, the cloning of the genes for ACC synthase and ACC oxidase has enabled scientists to use biotechnology to reduce ethylene production during fruit ripening, and thus to prolong the storage life of fruit. In the case of tomato, two different strategies have been employed to control ethylene production.

- The first strategy was to use the Ti plasmid of *Agrobacterium tumefaciens* to transform tomato tissue with antisense DNA for ACC synthase or ACC oxidase. When the genes for the two enzymes are expressed in the reverse direction in the cell, the resulting "antisense transcript" binds to and inactivates the "sense transcript" produced by the wild-type gene, effectively shutting down expression of the wild-type gene. After transformation, intact plants can be regenerated in tissue culture.
- In the second strategy for reducing ethylene production, tomato tissue was transformed (using *A. tumefaciens*) with genes encoding enzymes that metabolize ACC or AdoMet. In this case, the ACC that is produced is rapidly degraded, and no ethylene can be synthesized.

Regardless of the strategy used, all fruits from these genetically engineered plants show significant delays in ripening. Tomato fruits from plants transformed with the ACC synthase antisense DNA fail to ripen unless treated with ethylene.

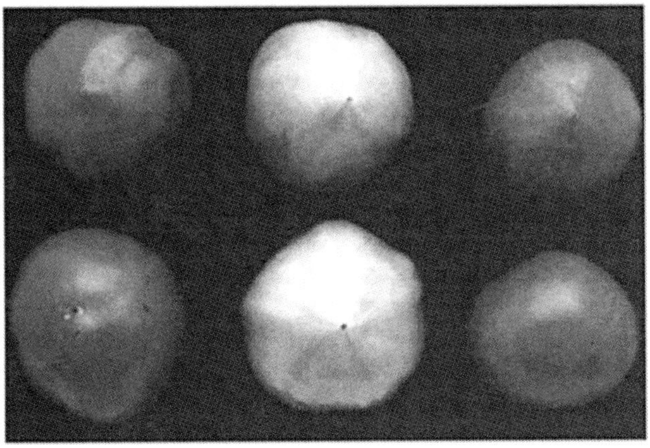

Figure: *Inhibition of tomato fruit ripening in detached tomato fruits by antisense ACC synthase RNA. Antisense tomato fruits (centre) remain firm and green unless treated by ethylene. Ethylene-treated antisense fruits (left) are indistinguishable from naturally ripened fruits (right) in terms of colour, flavour, aroma, and texture. (Reprinted with permission from Oeller et al. 1991.)*

Similar approaches are now being carried out in other agronomically important crops, such as melon and banana. In addition, these approaches are being used to retard floral senescence in carnations, thereby reducing postharvest losses. The research on ACC synthase and other ethylene biosynthesis genes thus provides an excellent example of how basic research can lead to improvements in agricultural and horticultural crops, the full benefits of which we are only beginning to realize.

7

Regulation of Transcription by Cis-Acting Sequences

The precise number and type of *cis*-acting sequences involved in regulating transcription of a given gene have been determined by various approaches:

Conserved sequences. The sequences of the promoter regions of a given gene (for example, *RBCS*) are first compared within members of the gene family in a particular species, and between gene family members of different species. Sequences acting as *cis*-acting regulatory elements are often represented in many species (that is, they are conserved) because they are essential for function, while the DNA sequences around them may be less essential, and therefore more variable. This approach gives an idea of the locations of sequences that may be involved in regulating the expression of a gene. For *RBCS* genes, the regulatory element typically encompasses a region about 1 kilobase of DNA upstream from the transcriptional start site.

Identification of response elements. Once general locations of the conserved sequences have been identified, reporter constructs of the type already described can be tested to determine whether this fragment is necessary and sufficient to confer light regulation on the expression of the reporter sequence. Then increasing amounts of the 52 end of the promoter region can be deleted to determine the minimum piece of DNA required for expression. Typically, 250 base pairs of DNA upstream of the transcription start site are required for light-regulated transcription. Internal deletions or the insertion of small, unrelated sequences, called *linkers*, can be used to define *cis*-acting regulatory elements.

Conferring light regulation. Synthetic copies of putative elements that contain one or more copies of a sequence can be inserted into non-light-regulated promoters to see if they can confer light regulation.

Identification of *trans*-acting factors. In vitro experiments in which nuclear protein extracts are allowed to bind to these different DNA fragments can be used to characterize the protein factors that bind to the *cis*-acting elements. Together, these four approaches have yielded detailed promoter maps for several genes, including *RBCS* and oat *PHYA*. Some of the important *cis*-acting elements present in many light-regulated genes include GT-1 sites (GGTTAA), I boxes (GATAA) and G boxes (CACGTC).

The *trans*-acting factors that bind to these three elements have also been characterized. However, each of these three *cis*-acting regulatory elements has also been found in the promoters of some non-light-regulated genes. Moreover, none of these elements alone is able to confer light regulation onto a non-light-regulated gene. Thus, light regulation arises from various combinations of different general elements rather than from a single regulatory element. Recent evidence suggests that the minimal sequence required for light-inducible gene expression is a pairwise combination of two elements such as GT-1 sites, G boxes, or I boxes.

Phytochrome Regulation of Gene Expression

We have seen that phytochrome-mediated light responses at the cellular level can involve rapid effects on membrane properties. The stimulation and repression of transcription by light can also be very rapid, with lag times as short as 5 minutes. Such early-gene expression is likely to be regulated by the direct interaction of transcription factors by one or more phytochrome-initiated signal transduction pathways. Within the nucleus, these transcription factors stimulate the transcription of specific genes. Some of the gene products of the early genes are transcription factors themselves, which activate the expression of other genes. Expression of the early genes, also called primary response genes, is independent of protein synthesis; expression of the late genes, or secondary response genes, requires the synthesis of new proteins.

As with cytokinin-regulated gene expression, the photoregulation of gene expression has focused on the nuclear genes that encode messages for chloroplast proteins: the small subunit (SSU) of ribulose-1,6-bisphosphate carboxylase/oxygenase (rubisco) and the major light-harvesting chlorophyll *a/b*–binding proteins associated with the light-harvesting complex of photosystem II (LHCIIb proteins). These proteins play important roles in chloroplast development and greening; hence their regulation by phytochrome has been studied in detail. The genes for both of these proteins—*RBCS* and *LHCB* (also called *CAB* in some studies)—are present in multiple copies in the genome.

The abundance of these mRNAs has been shown to increase in response to light and phytochrome action. The abundance of other mRNAs, however, can be decreased by phytochrome action; this latter class includes mRNAs that encode oat phyA and the chlorophyll biosynthesis enzyme NADPH–protochlorophyllide oxidoreductase. Although the abundance of most of these mRNAs is regulated by light given in the low-fluence range, some *LHCB* mRNAs also respond to light in the very-low-fluence range.

We can demonstrate phytochrome-regulated induction of mRNA abundance (for example *RBCS* mRNAs) experimentally by giving etiolated plants a brief pulse of low-fluence red or far-red light, returning them to darkness to allow the signal transduction pathway to operate, and then measuring the abundance of specific mRNAs in total RNA prepared from each set of plants. If its abundance is regulated by phytochrome, the mRNA is absent or present at low levels in etiolated plants, but is increased by red light. The red light–induced increase in expression can be reversed by immediate treatment with far-red light, but far-red light alone has little effect on mRNA abundance.

In the case of mRNAs (e.g., *LHCB*) that respond to very-low-fluence red light, expression increases in response to both red and far-red light, and additional experiments are needed to confirm the involvement of phytochrome. In the case of mRNAs whose expression is negatively regulated by phytochrome, for example oat *PHYA* mRNA, levels are *high* in darkness, *reduced* by a pulse of red light, and *restored* to dark levels by red/far-red light treatment. However, none of these experiments prove the action of phytochrome on transcription; instead they show the effect of phytochrome only on overall mRNA abundance in total RNA, which reflects both transcription and degradation processes.

As mentioned earlier, the abundance of oat *PHYA* mRNA in total RNA is the result of both transcription and degradation events.

The involvement of phytochrome in transcriptional regulation of mRNA abundance has been demonstrated by two approaches. In the first approach, etiolated plants are treated as was described for the mRNA studies, and then transcriptionally active nuclei are isolated and allowed to elongate transcripts initiated under the different in vivo light treatments in the presence of radioactive precursors. If the relative incorporation into an mRNA is similar to that described for the overall abundance in total RNA, then transcription of the mRNA is thought to be under phytochrome control. The second approach involves fusing the promoter of the gene to be tested with the coding sequence of a reporter gene such as *GUS*. By placing the fused gene back into the plant it is possible to measure the abundance of the reporter mRNA or its encoded enzyme activity. Both of these approaches have been used extensively, and the latter has allowed the definition of the minimum amount of the upstream promoter region required to confer phytochrome-regulated transcription of the linked reporter sequence.

Recently, Elaine Tobin and her colleagues at the University of California, Los Angeles, characterized a MYB-related transcription factor whose mRNA level increases rapidly when *Arabidopsis* is transferred from the dark to the light. This DNA-binding protein appears to bind to the promoter of an *LHCB* gene and regulate its transcription,, occurs later than the increase in the MYB related protein. The gene that encodes the MYB-related protein is therefore probably a primary response gene, while the *LHCB* gene is probably a secondary response gene.

More recently, microarray analysis has been applied to the study of light signal transduction. In microarray analysis short oligonucleotides or cDNA sequences representing thousands of gene sequences are spotted on a glass slide. RNA samples isolated from control and experimental plants are labelled using two different fluorescent dyes. Both the control and the experimental sample are hybridized to the microarray and the slide is then scanned using two lasers that will excite one or the other dye. The light emitted from the dye is proportional to the amount of RNA that is bound to the gene-specific oligonucleotide or cDNA sequence and thus provides a relative measure of gene expression. To visualize the relative expression levels from both samples, images are often false coloured so that one signal is represented with green and the other signal as a red, when the signals overlap they are yellow. The commercially available Affymetrix arrays utilizes small 25-mer oligos as probes on the glass slide that are hybridized to a single labelled sample. This technique has been used to examine changes in gene expression when plants are grown under red or far-red light to define phytochrome signalling pathways and have shown that multiple transcription factors are early targets of phytochrome control.

Two-Hybrid Screens and Co-Immunoprecipitation

Two-hybrid library screens are based on the finding that many transcriptional activators contain separable DNA binding and transcriptional activation domains. In a two-hybrid screen, a DNA binding domain of a transcription factor is fused to a test or "bait" protein such as phyA. Yeast cells carrying the bait protein are transformed with a library of plant cDNA products that are fused to a transcriptional activation domain ("prey"). If the bait protein interacts with the prey protein, then the activation domain will be brought into close proximity to the DNA binding site of the transcriptional activator, thus driving reporter gene expression that will pigment the yeast cells or provide an essential amino acid, permitting growth.

Co-immunoprecipitation experiments are often conducted to provide independent verification of the results of two-hybrid screens. These assays can be performed using *in vitro* synthesized versions of bait and prey proteins. The proteins are mixed together and immunoprecipitated using an antibody that is specific to either the bait or prey construct. If the two proteins co-precipitate, it is evidence they interact with each other.

The Roles of G-Proteins and Calcium in Phytochrome Responses

Early experiments with intact seedlings and cultured cells indicated that the control of gene expression by phytochrome involves the action of a heterotrimeric G-protein and the calcium-binding regulatory protein calmodulin.

Subsequently, a single-cell system has been developed that can be microinjected with various proposed components of the pathway and monitored for changes in gene expression. Although these experiments have not yet led to the identification of specific pathway components, they have been informative about the involvement of G-proteins and calmodulin.

Microinjection of phytochrome (mainly phyA) purified from etiolated oat seedlings into single cells prepared from the tomato *aurea* mutant (a chromophore biosynthetic mutant that has reduced levels of all phytochromes) results in stimulation of chloroplast development, accumulation of anthocyanin pigment, and expression of a GUS-coding sequence fused to a phytochrome-regulated promoter sequence. Etiolated seedlings of the *aurea* mutant show poor induction of these responses when transferred to light, probably because of the reduction in the levels of functional phyA. Inhibitors of G-proteins block the phytochrome-stimulated events in the single-cell system, while activators or intermediates stimulate the response. Increased levels of Ca^{2+} in the medium also stimulate *GUS* expression and chloroplast development when calmodulin is present, but they do not affect anthocyanin biosynthesis.

In the *Aurea* cell system, cGMP has been shown to act as a signalling intermediate in at least two signal transduction pathways. Specifically, the pathway that activates anthocyanin biosynthesis requires cGMP. Inhibitors of cGMP action inhibit this pathway; addition of cGMP activates it. In contrast, the pathway that requires calcium to activate chloroplast development and gene expression has a variable requirement for cGMP. Activation of genes that encode several chloroplast proteins, including RBCS and LHCB, does not require cGMP, while activation of genes that encode some photosystem I polypeptides and the cytochrome *b6/f* complex does.

The difference in signalling requirements is due in part to the different *cis*-acting regulatory elements found upstream of these genes. Specific DNA elements upstream of *RBCS* genes appear to be the targets for the calcium pathway, while other distinct elements located upstream of anthocyanin biosynthesis genes are the targets for cGMP. The differential regulation of the expression of phytochrome-regulated genes can now be explained in terms of their *cis*-acting regulatory elements, different combinations of which afford both tissue-specific and light-specific expression.

Phytochrome Effects on Ion Fluxes

Phytochrome can rapidly alter the properties of membranes. We have already seen that low-fluence red light is required before the dark period to induce rapid leaflet closure during nyctinasty, and that fluxes of K^+ and Cl^- into and out of extensor and flexor cells mediate the response. However, the rapidity of leaf closure in the dark (lag time about 5 minutes) would seem to rule out mechanisms based on gene expression. Instead, rapid phytochrome-induced changes in membrane permeability and transport appear to be involved.

The role of the plasma membrane proton pump in *Samanea* pulvini (the structures at the base of the petiole that include the extensor and flexor cells) was studied by use of a proton-sensitive liquid membrane electrode. During phytochrome-mediated leaflet closure, the apoplastic pH of the flexor cells (the cells that swell during leaflet closure) decreased, while the apoplastic pH of the extensor cells (the cells that shrink during leaflet closure) increased. Thus, the plasma membrane H^+ pump of the flexor cells appears to be activated by darkness (provided that phytochrome is in the Pfr form), whereas the H^+ pump of the extensor cells appears to be deactivated under the same conditions.

Increased H^+ pump activity on the plasma membranes of the flexor cells should increase the membrane potential, thereby allowing more K^+ to enter the cells. Conversely, a reduction in the plasma membrane H^+ pump activity in the extensor cells would promote the loss of K^+ ions. Consistent with this model, the reverse pattern of apoplastic pH change was seen during leaflet opening.

Studies have also been carried out on phytochrome regulation of K^+ channels. Since it is difficult to study membrane channels in the intact pulvinus, Richard Crain and his colleagues at the University of Connecticut isolated protoplasts (cells without their cell walls) of both extensor and flexor cells from *Samanea* leaves. They determined the open or closed state of the K+ channels indirectly by measuring the membrane potential changes that occurred when the extracellular K^+ concentration was raised from 20 to 200 m*M*. When the extracellular K^+ concentration was raised, K^+ entered the protoplasts and depolarized the membrane potential only if the K^+ channels were open. When the extensor and flexor protoplasts were transferred to constant darkness, the state of the

K^+ channels exhibited a circadian rhythmicity during a 21-hour incubation period and the two cell types varied reciprocally, just as they do in vivo. That is, when the extensor K^+ channels were open, the flexor K^+ channels were closed, and vice versa. Thus, the circadian rhythm of leaf movements has its origins in the circadian rhythm of K^+ channel opening.

To test the effects of phytochrome on K^+ channels, researchers first treated protoplasts with white light, then exposed them to a brief pulse of either red light alone or red light followed by far-red light and transferred them to continuous darkness. Red light caused the K^+ channels of the flexor protoplasts to open, and far-red light reversed the effect of red light. The extensor protoplasts were closed by both treatments. In contrast to red light, blue light caused the K^+ channels of the extensors to open, and those of the flexors to close. Thus, phytochrome seems to regulate specifically the K^+ channels of the flexor cells. Blue light regulates primarily the extensor cells, although it can also affect the flexors at specific times of day.

On the basis of the evidence thus far, we can conclude that phytochrome brings about leaflet closure by regulating the activities of the primary proton pumps and/or the K+ channels of the flexor and extensor cells. Although the effect is rapid, it is not instantaneous, and it is therefore unlikely to be due to a direct effect of phytochrome on the membrane. Instead, phytochrome acts indirectly via one or more signal transduction pathways, as in the case of the regulation of gene expression by phytochrome. However, some effects of red and far-red light on the membrane potential are so rapid that phytochrome may also interact directly with the membrane. Such rapid modulation has been measured in individual cells and has been inferred from the effects of red and far-red light on the surface potential of roots and oat (*Avena*) coleoptiles. The lag between the production of Pfr and the onset of measurable potential changes varies from organism to organism; for example, it is about 1.7 s for depolarization (inside becomes less negative relative to outside the plasma membrane) in the giant alga *Nitella* and 4.5 s for hyperpolarization in *Avena*.

Changes in the bioelectric potential of cells imply changes in the flux of ions across the plasma membrane. The *Mougeotia* microbeam irradiation experiments of Haupt seemed to support the idea that phytochrome is localized on the plasma membrane, and membrane isolation studies carried out in the 1970s provided evidence that a small portion of the total phytochrome is tightly bound to various organellar membranes.

These findings led some workers to suggest that membrane-bound phytochrome represents the physiologically active fraction, and that all the effects of phytochrome on gene expression are initiated by changes in membrane permeability. On the basis of sequence analysis, however, it is now clear that phytochrome is a hydrophilic protein without membrane-spanning domains. The current view is that phytochrome in *Mougeotia* is associated with microtubules located directly beneath the plasma membrane.

If phytochrome exerts its effects on membranes from some distance, no matter how small, involvement of a "second messenger" would be implied. For example, rapid changes in cytosolic free calcium have been implicated as second messengers in several signal transduction pathways, and there is evidence that calcium plays a role in chloroplast rotation in *Mougeotia*.

More recently, however, the role of calcium as a second messenger in phytochrome action has been questioned, so calcium-independent mechanisms may also be involved.

Microarray Analysis of Shade Avoidance

To identify components of the shade-avoidance response pathway, microarray analysis was conducted in Arabidopsis, comparing plants grown in low and high R:FR light environments. These studies have identified both early-and late-acting genes that regulate responses to shading by other plants.

Characterizations of two Arabidopsis transcription factors (*ATHB-2* and *PIL1*) also suggests that shade avoidance is initiated by rapid transcriptional changes following a decrease in the ratio of red to far-red light.

These profiling studies strengthen the previous genetic studies that show that phyA acts to antagonize phyB-mediated responses under prolonged low R:FR light conditions through the transcriptional control of target genes that include components of the auxin transport and signalling pathways. Interestingly, it appears that the circadian clock is capable of regulating or "gating" physiological responses to R:FR through a modulation of gene expression. That is, changes in gene expression due to a change in the R:FR only occur at certain phases of the circadian rhythm.

Blue-Light Responses: Stomatal Movements and Morphogenesis

Guard Cell Osmoregulation and a Blue Light-Activated Metabolic Switch

Studies on photosynthetic algae, such as *Chlorella,* have shown that irradiation by blue light increases respiratory rates of dark-grown cells, with an action spectrum that is typical of specific blue-light responses. These changes in respiratory rates are associated with a drastic shift in metabolism, which can also be observed in experiments using blue and red light.

Red light stimulates photosynthetic carbon fixation in these algae, and the fixed carbon is incorporated into glucose and sucrose. The metabolic shift is observed when low photon fluxes of blue light are added to the red-light treatment; glucose and sucrose are no longer synthesized, and the fixed carbon is used in the biosynthesis of organic acids, amino acids, and proteins.

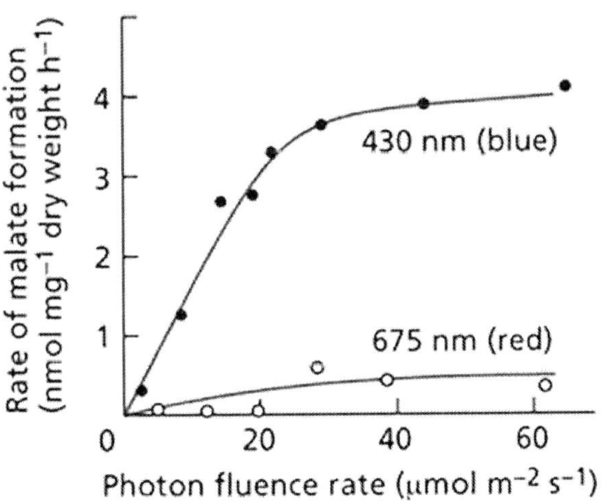

Figure: *Blue light stimulates malate synthesis in guard cells. Under red light, guard cells synthesize mostly glucose and sucrose; this metabolic pathway is deactivated by blue light, and metabolism switches to starch degradation and malate biosynthesis.*

Blue light is also a powerful regulator of guard cell metabolism. Guard cells irradiated with red light accumulate sucrose; if low photon fluxes of blue light are added to the red-light irradiation, sucrose stops accumulating, and guard cells accumulate K^+ and malate^{2-}. Thus, the blue light–stimulated shift to the synthesis of the organic acid malate in guard cells parallels the shift to the synthesis of organic acids and amino acids in photosynthetic algae.

The experimental use of blue and red light has been useful for investigations of osmoregulatory pathways associated with stomatal movements. The rapid activation of proton pumping by blue light generates a proton gradient at the guard cell plasma membrane that drives potassium and chloride uptake. Blue light also stimulates starch hydrolysis and malate biosynthesis. Prolonged incubations of guard cells in detached epidermis under low fluence rates of blue light result in large stomatal apertures accompanied by extensive starch hydrolysis.

Guard cells also show a sucrose-dependent osmoregulatory phase. Sucrose can originate from starch hydrolysis. On the other hand, guard cells from detached epidermis irradiated with red light under ambient CO_2 concentrations operate a third metabolic pathway, showing no detectable potassium uptake or starch hydrolysis. In these conditions, guard cells slowly accumulate sucrose, and this sucrose accumulation is blocked by DCMU (dichlorophenyldimethylurea), an inhibitor of photosynthetic electron transport. A recent study of guard cell osmoregulation under red light illumination and CO_2-free air showed that under these conditions, in which photosynthesis is inhibited by the absence of CO_2, red light activates the potassium-dependent pathway. These studies underscore the remarkable metabolic flexibility of guard cells. A study of the effects of calcium and abscisic acid on stomatal apertures under red and blue light found that increasing calcium or ABA concentrations inhibited blue light–stimulated stomatal opening in a concentration-dependent fashion, but had no effect on red light–stimulated opening. These contrasting responses to blue and red light can be explained by the effect of the treatments on guard cell osmoregulation.

ABA and high calcium concentrations have been shown to inhibit proton pumping and potassium uptake, which are central to blue light–stimulated stomatal opening. Red light, on the other hand, stimulates guard cell photosynthesis and sucrose accumulation, and this osmoregulatory pathway appears to be insensitive to ABA and calcium. Unlike visual systems, the colour information in the blue light and red light treatments does not appear to have functional significance for photosynthetic organisms. Rather, their dramatic experimental effects appear to be a result of a differential stimulation of carotenoids and chlorophylls that are selectively excited by blue light applied under a saturating background of red light, or by red light alone, respectively. Under solar radiation, regulatory pathways activated by the excitation of these pigments are used by the chloroplast to sense key features in their environment and to control their metabolism.

The Coleoptile Chloroplast

Guard cell chloroplasts are a conspicuous structural feature of guard cells, and their physiology and biochemistry have been studied for several decades. In contrast, the chloroplasts of phototropically sensitive coleoptiles have only recently been analysed in detail. The coleoptile chloroplast had been suggested as a possible site for photoreception in oat coleoptiles by Kenneth V. Thimann and his coworkers in 1960, but interest in the flavin hypothesis later in that decade shifted attention away from coleoptile chloroplasts and these organelles have been virtually ignored for several decades.

Recent studies have shown that the coleoptile chloroplast evolves oxygen, fixes CO_2 photosynthetically, and operates the xanthophyll cycle. Dark-grown, etiolated coleoptile tips have violaxanthin and antheraxanthin but no zeaxanthin, and they accumulate zeaxanthin when irradiated.

When dark-grown coleoptiles are pretreated with red light for different lengths of time, their zeaxanthin content is proportional to the length of the red-light treatment. Exposure of coleoptiles with different zeaxanthin content to a brief pulse of blue light elicitis phototropic bending that is linearly related to the zeaxanthin content of the coleoptiles. Most interestingly, the regression line through the data points extrapolates to zero, suggesting that there is no phototropic response in the absence of zeaxanthin.

Red light–treated coleoptiles with a high zeaxanthin content convert their zeaxanthin to violaxanthin when transferred to the dark. Phototropic bending in response to a pulse of blue light given at different time intervals after the transfer to the dark is linearly related to zeaxanthin content.

When corn coleoptiles are given a saturating red-light pretreatment and a blue-light pulse in the presence of increasing concentrations of DTT (an inhibitor of zeaxanthin formation), bending and zeaxanthin content decrease linearly as a function of DTT concentration.

Thus the relationship between the zeaxanthin content of coleoptile chloroplasts and blue light–dependent phototropic sensitivity is very similar to the relationship between the zeaxanthin content of the guard cell chloroplast and the blue-light sensitivity of stomatal opening.

Coleoptile and guard cell chloroplasts share some properties that are clearly distinct from mesophyll chloroplasts, including high rates of oxygen evolution and low rates of photosynthetic carbon fixation, a high sensitivity of their xanthophyll cycle to low levels of incident photon fluxes, and a high starch content. The coleoptile and guard cell chloroplasts also show a specific blue light response with an action spectrum that closely matches the action spectra for blue light–stimulated stomatal opening and coleoptile phototropism.

These remarkable similarities suggest that these distinct characteristics are typical of chloroplasts specialized in signal transduction, as opposed to the classic role of the mesophyll chloroplast, specialized in photosynthetic carbon fixation.

8

Principles of Spectrophotometry

Much of what we know about the photosynthetic apparatus was learned through spectroscopy—that is, measurements of the interaction of light and molecules. Spectrophotometry is an important branch of spectroscopy that focuses on the technique of measurement. Here we will examine four topics: Beer's law, the measurement of absorbance, action spectra, and difference spectra.

Beer's Law

An essential piece of information about any molecular species is how much of it is present. Quantitative measures of concentration are one of the cornerstones of biological science. Of all the methods that have been devised for measuring concentration, by far the most widely applied is absorption spectrophotometry. In this technique, the amount of light that a sample absorbs at a particular wavelength is measured and used to determine the concentration of the sample by comparison with appropriate standards or reference data. The most useful measure of light absorption is the absorbance (A), also commonly called the optical density (OD). The absorbance is defined as $A = \log I_0 / I$ where I_0 is the intensity of light that is incident on the sample and I is the intensity of light that is transmitted by the sample.

The absorbance of a sample can be related to the concentration of the absorbing species through Beer's law:

$$A = \varepsilon cl$$

where *c* is concentration, usually measured in moles per litter; *l* is the length of the light path, usually 1 cm; and e is a proportionality constant known as the molar extinction coefficient, with the units of litres per mole per centimetre.

The value of e is a function of both the particular compound being measured and the wavelength. Chlorophylls typically have an e value of about 100,000 L mol^{-1} cm^{-1}. When more than one component of a complex mixture absorbs at a given wavelength, the absorbances due to the individual components are generally additive.

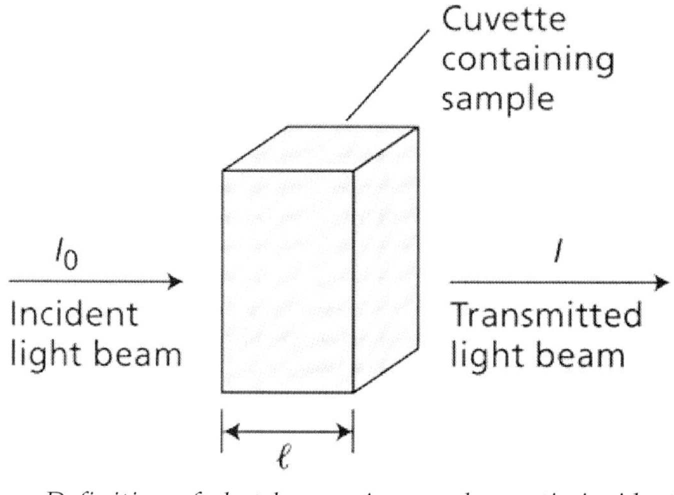

Figure: Definition of absorbance. A monochromatic incident light beam of intensity I_0 traverses a sample contained in a cuvette of length (l). Some of the light is absorbed by the chromophores in the sample, and the intensity of light that emerges is I.

The Spectrophotometer

The absorbance is measured by an instrument called a spectrophotometer. The essential parts of a spectrophotometer include a light source, a wavelength selection device such as a monochromator or filter, a sample chamber, a light detector, and a readout device, usually also include a computer, which is used for storage and analysis of the spectra. The most useful machines scan the wavelength of the light that is incident on the sample and produce, as output, spectra of absorbance versus wavelength.

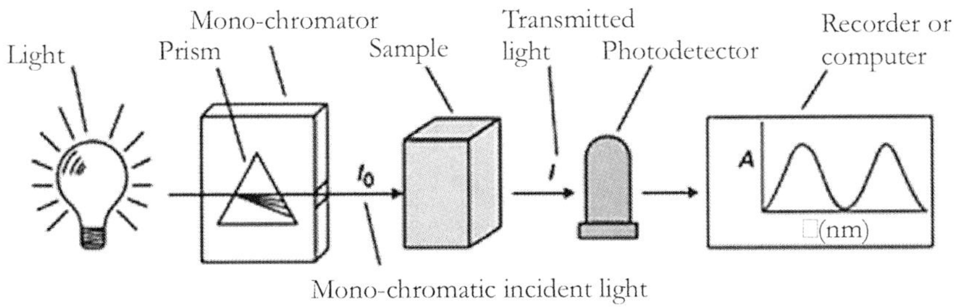

Figure: *Schematic diagram of a spectrophotometer. The instrument consists of a light source, a monochromator that contains a wavelength selection device such as a prism, a sample holder, a photodetector, and a recorder or computer. The output wavelength of the monochromator can be changed by rotation of the prism;*
the graph of absorbance versus wavelength is called a spectrum.

Action Spectra

The use of action spectra has been central to the development of our current understanding of photosynthesis. An action spectrum is a graph of the magnitude of the biological effect observed as a function of wavelength. Examples of effects measured by action spectra are oxygen evolution and hormonal growth responses due to the action of phytochrome. Often an action spectrum can identify the chromophore responsible for a particular light-induced phenomenon. Action spectra were instrumental in the discovery of the existence of the two photosystems in O_2-evolving photosynthetic organisms.

Some of the first action spectra were measured by T. W. Engelmann in the late 1800s. Engelmann used a prism to disperse sunlight into a rainbow that was allowed to fall on an aquatic algal filament. A population of O_2-seeking bacteria was introduced into the system. The bacteria congregated in the regions of the filaments that evolved the most O_2. These were the regions illuminated by blue light and red light, which are strongly absorbed by chlorophyll. Today, action spectra can be measured in room-sized spectrographs in which the scientist enters a huge monochromator and places samples for irradiation in a large area of the room bathed by monochromatic light. But the principle of the experiment is the same as that of Engelmann's experiments.

Schematic diagram of the action spectrum measurements by T. W. Engelmann. Engelmann projected a spectrum of light onto the spiral chloroplast of the filamentous green alga *Spirogyra* and observed that oxygen-seeking bacteria introduced into the system collected in the region of the spectrum where chlorophyll pigments absorb. This action spectrum gave the first indication of the effectiveness of light absorbed by accessory pigments in driving photosynthesis.

Difference Spectra

An important technique in studies of photosynthesis is light-induced difference spectroscopy, which measures changes in absorbance. In this technique, bright light, often called actinic light, is used to illuminate a sample, while a dim beam of light is used to measure the absorbance of the sample at wavelengths other than that of the actinic beam. In this way a difference spectrum is obtained, which represents the changes in the absorption spectrum of the sample induced by illumination with the actinic light. Absorption bands that disappear upon illumination appear as negative peaks; new bands that appear upon illumination appear as positive peaks. Difference spectra give important clues to the identity of molecular species participating in the photoreactions of photosynthesis.

By the use of special flash techniques, it is possible to record the difference spectrum at a given time after flash excitation. Multiple difference spectra recorded at different times after flash excitation can be used to measure the kinetics of the chemical reactions that follow photon excitation of a reaction centre. These techniques can have extraordinary time resolution, in some cases less than a picosecond (10^{-12} s), and have provided great insights into the earliest events in the photosynthetic energy storage process.

Antagonistic Effects of Light on Cytochrome Oxidation

The puzzling red drop and enhancement effects were explained by experiments performed by Louis Duysens of the Netherlands. He suggested that two phtotchemical events were responsible for enhancement effects. One photochemical event produced an oxidation while the other produced a reduction. Chloroplasts contain cytochromes, iron-containing proteins that function as intermediate electron carriers in photosynthesis. Duysens found that when a sample of a red alga was illuminated with long-wavelength light, the cytochrome became mostly oxidized. If light of a shorter wavelength was also present, the effect was partly reversed. These antagonistic effects can be explained by a mechanism involving two photochemical events: one that tended to oxidize the cytochrome and one that tended to reduce it.

We know now that in the red region of the spectrum, one of the photoreactions, known as photosystem I (PS-I), absorbs preferentially far-red light of wavelengths greater than 680 nm, while the second, known as photosystem II (PS-II), absorbs red light of 680 nm well and is driven very poorly by far-red light. This wavelength dependence explains the enhancement effect and the red drop effect. Another difference between the photosystems is that Photosystem I produces a strong reductant, capable of reducing $NADP^+$, and a weak oxidant. Photosystem II produces a very strong oxidant, capable of oxidizing water, and a weaker reductant than the one produced by photosystem I. This reductant of photosystem II rereduces the oxidant produced by photosystem I, which explains the antagonistic effect.

Structures of Two Bacterial Reaction Centres

In 1984, Hartmut Michel, Johann Deisenhofer, Robert Huber, and coworkers in Munich solved the three-dimensional structure of the reaction centre from the purple photosynthetic bacterium *Rhodopseudomonas viridis*. This landmark achievement, for which a Nobel prize was awarded in 1988, was the first high-resolution X-ray structure determination for an integral membrane protein and the first structure determination for a reaction centre complex.

The protein part of the complex consists of four separate polypeptides. Two of them, called L and M (for *l*ight and *m*edium mass) bind all of the bacteriochlorophyll, quinone, and carotenoid cofactors of the complex. The structure has a twofold symmetry about an axis perpendicular to the plane of the membrane, hinting at a dimeric nature of the reaction centre. The ten transmembrane portions of the L and M peptides (five from each) are arranged in α helices, and there are almost no charged amino acid residues in the interior of the membrane. The H (*h*eavy) protein has a single transmembrane helix and is localized mostly on the cytoplasmic side of the membrane. The C (*c*ytochrome) subunit is located in the periplasmic region (the region between the bacterial plasma membrane and the outer membrane). The geometric arrangement of the pigments and the quinones (electron acceptors), with the protein removed. A similar arrangement is found in the reaction centre of another purple photosynthetic bacterium, *Rhodobacter sphaeroides*, except that the C subunit is not present.

Two of the bacteriochlorophyll molecules are in intimate contact with each other and are known as the special pair. This dimer, whose existence was predicted from magnetic-resonance studies, is the photoactive portion of the complex. An electron is transferred from this dimer along the sequence of electron carriers on the right side of the complex. Detailed analysis of these structures, along with analysis of numerous mutants, has revealed many of the principles involved in the energy storage processes that are carried out by all reaction centres.

The bacterial reaction centre structure is thought to be similar in many ways to that found in photosystem II from oxygen-evolving organisms, especially in the electron acceptor portion of the chain. The proteins that make up the core of the bacterial reaction centre are relatively similar in sequence to their photosystem II counterparts, implying an evolutionary relatedness.

Midpoint Potentials and Redox Reactions

Redox reactions, midpoint potentials, and their relationship to the laws of thermodynamics. These concepts are useful for our discussion of electron flow from H_2O to $NADP^+$ and the interactions between the different electron carriers.

The midpoint potential (E_m) is a measure of the tendency of a compound to take electrons from other compounds. A large positive midpoint potential means that the compound is a strong oxidant; a large negative value means that the compound is a strong reductant (in both cases relative to the standard hydrogen electrode).

Equilibrium constants can easily be predicted from midpoint potentials, in the same way that free energies were related to equilibrium constants. Midpoint potentials for many chemical and biochemical reactions have been measured and tabulated. The y-axis on the Z scheme, midpoint potentials of the electron carriers, with negative values higher than positive ones. This choice makes reactions that are spontaneous (releasing free energy) appear "downhill" on the graph.

Knowledge of the midpoint potentials of the various electron carriers is important in establishing the pathway of electron flow in any biochemical electron transport system, such as those found in chloroplasts or mitochondria. Researchers make this measurement usually by carrying out a redox titration. They adjust, or poise, the sample at a particular redox potential, usually by adding small amounts of oxidants or reductants. Redox mediators, small molecules that permit rapid equilibration between the sample and the electrodes of the measurement system, must be included to ensure that the system is at equilibrium when the measurement is made. Several measurements are made at a variety of redox potentials. The sample is stirred in a special cell that contains platinum and reference electrodes, and chemical oxidants and reductants are added to adjust the redox potential, which is read with a voltmeter.

We can measure the extent of the redox reaction by following a particular property of the sample, usually absorbance, at each potential. In the example illustrated, the reduced form of the compound has an absorbance that decreases as the compound is oxidized. The fraction of reduced form at each potential is plotted against redox potential, and the midpoint potential (E_m) is determined as the potential at which the compound is half oxidized and half reduced.

Oxygen Evolution

The chemical mechanism of photosynthetic water oxidation is not yet known, although there is a great deal of indirect evidence about the process. If a sample of dark-adapted photosynthetic membrane is exposed to a sequence of very brief, intense flashes, a characteristic pattern of oxygen production is observed. Little or no oxygen is produced on the first two flashes, and maximal oxygen is released on the third flash and every fourth flash thereafter, until eventually the yield per flash damps to a constant value. This remarkable result was first observed by Pierre Joliot in the 1960s.

A schematic model explaining these observations, proposed by Kok and coworkers, has been widely accepted. This model for the photooxidation of water, called the S state mechanism, consists of a series of five states, known as S_0 to S_4, which represent successively more oxidized forms of the water-oxidizing enzyme system, or oxygen-evolving complex. The light flashes advance the system from one S state to the next, until state S_4 is reached.

State S_4 produces O_2 without further light input and returns the system to S_0. Occasionally, a centre does not advance to the next S state upon flash excitation, and less frequently, a centre is activated twice by a single flash. These misses and double hits cause the synchrony achieved by dark adaptation to be lost and the oxygen yield eventually to damp to a constant value. After this steady state has been reached, a complex has the same probability of being in any of the states S_0 to S_3 (S_4 is unstable and occurs only transiently), and the yield of O_2 becomes constant. States S_2 and S_3 decay in the dark, but only as far back as S_1, which is stable in the dark. Therefore, after adaptation to the dark, approximately three-fourths of the oxygen-evolving complexes appear to be in state S_1 and one-fourth in state S_0. This distribution of states explains why the maximum yield of O_2 is observed after the third of a series of flashes given to dark-adapted chloroplasts.

Photosystem I

The PS-I reaction centre is composed of a multiprotein complex. The reaction centre chlorophyll P700 and about 100 core antenna chlorophylls are bound to two proteins, PsaA and PsaB, with molecular masses in the range of 66 to 70 kDa. PS-I reaction centre complexes have been isolated from several organisms and found to contain the 66 to 70 kDa proteins, along with a variable number of smaller proteins in the range of 4 to 25 kDa. Some of these proteins serve as binding sites for the soluble electron carriers plastocyanin and ferredoxin. The functions of some of the other proteins are not well understood. An 8-kDa protein contains some of the bound iron-sulfur centres that serve as early electron acceptors in photosystem I. The structure of the PS-I complex from pea has been determined to a resolution of 4.4 E, and the positions of many of the chlorophylls and electron transfer components have been located.

In their reduced form, the electron carriers that function in the acceptor region of photosystem I are all extremely strong reducing agents. These reduced species are very unstable and thus difficult to identify. Evidence indicates that one of these early acceptors is a chlorophyll molecule, and another is a quinone species, phylloquinone, also known as vitamin K_1.

Additional electron acceptors include a series of three membrane-associated iron–sulfur proteins, or bound ferredoxins, also known as Fe–S centres Fe–S_X, Fe–S_A, and Fe–S_B. Fe–S centre Fe–S_X is part of the P700-binding protein; centres Fe–S_A and Fe–S_B reside on an 8 kDa protein that is part of the PS-I reaction centre complex. Electrons are transferred through centres Fe–S_A and Fe–S_B to ferredoxin, a small, water-soluble iron–sulfur protein. The membrane-associated flavoprotein ferredoxin–NADP reductase (FNR) reduces $NADP^+$ to NADPH, thus completing the sequence of noncyclic electron transport that begins with the oxidation of water. Electron transport between PSII and PSI is mediated by plastohydroquinone, the cytochrome b_6f complex, and plastocyanin. The cytochrome b_6f complex contains two *b*-type hemes and one *c*-type heme.

A model for the organization of electron carriers in PS-I is shown in Figure below. The P700 dimer is located at the bottom of the structure and two symmetrical arms radiate from P700. Each arm includes an accessory chlorophyll *a* and another chlorophyll molecule tentatively identified as A_0. Other structures include the Fe-S centre F_X, and two other Fe-S centres, F_1 and F_2. The distance between the carriers is shown at the right.

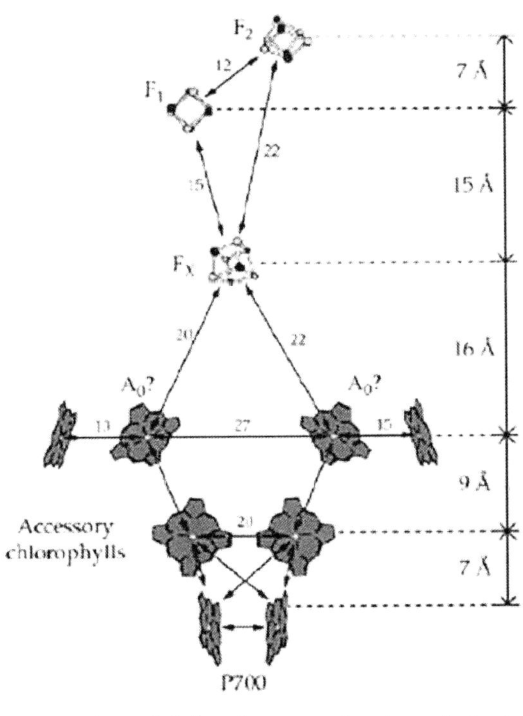

Figure: A model for the organization of electron carriers in PS-I.

The PS-I reaction centre appears to have some functional similarity to the reaction centre found in the anaerobic green sulfur bacteria and the heliobacteria.

These bacteria contain low-potential Fe–S centres as early electron acceptors and are probably capable of ferredoxin-mediated NAD^+ reduction similar to the $NADP^+$ reduction function of photosystem I. There is almost certainly an evolutionary relationship between these complexes and photosystem I of oxygen-evolving organisms.

ATP Synthase

The chemical mechanism by which ATP is synthesized is not yet understood in detail. However, considerable evidence now supports a mechanism, first proposed by Paul Boyer, in which the principal energy-requiring step is the release of bound ATP from the enzyme.

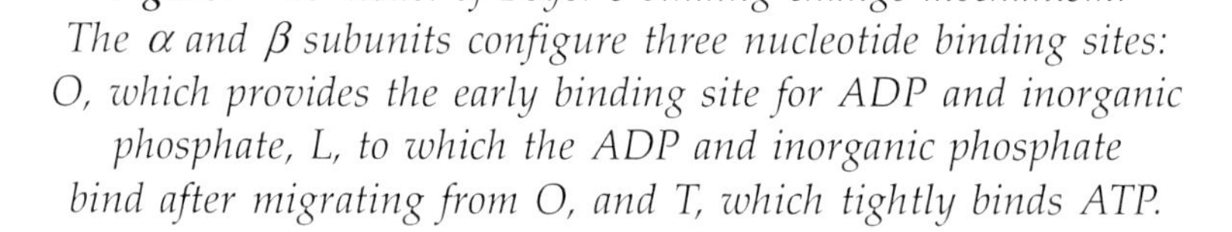

Figure: The model of Boyer's binding change mechanism. The α and β subunits configure three nucleotide binding sites: O, which provides the early binding site for ADP and inorganic phosphate, L, to which the ADP and inorganic phosphate bind after migrating from O, and T, which tightly binds ATP. Energy ensuing from the movement of protons from the chloroplast lumen to the stroma drives the rotation of the γ subunit of CF_1, and the interconversion of the binding sites and the release of an ATP molecule.

It has also been proposed that during catalysis a large portion of the CF_1 complex rotates about a bearing consisting of the γ subunit. The γ subunit may act as a camshaft does, rotating alternately against the α and α subunits. The energy of the conformational movements is then translated into phosphoanhydride bond energy.

The thylakoid ATP synthase has two segments: CF_0 is a transmembrane segment, and CF_1 is a hydrophilic segment at the stomal surface. CF_0 traslocates protons from the chloroplast lumen to the catalytic part of the enzyme and CF_1 cataylses the conversion of ADP and inorganic phosphate to ATP.

The chloroplast ATP synthase has nine different subunits. CF_1 is made of two large subunits, α and β, plus three smaller subunits, γ, δ, and ε. Each CF_1 molecule has three copies of the α and β subunits and one copy of the γ, δ and ε subunits. The α and β subunits bind ADP and phosphate and catalyize the phosphorylation of ADP into ATP.

Studies on the mitochondrial ATP synthase have increased our understanding of how the ATP synthases work. As proposed by Paul Boyer in 1997, ATP synthesis takes place by a binding change mechanism. According to this model, the energy of the proton gradient is used to release a tightly bound form of ATP from a catalytic binding side of the enzyme. As shown in Figure below, CF_1 has three nucleotide binding sites, L, T and O. ADP and inorganic phosphate are postulated to first bind to the O site, and then move to the L site. The T site binds ATP. As protons move from the lumen to the stromal region through the CF_0 channel, energy is released and the γ subunit of CF_1 rotates. The rotation causes conformation changes in the three nucleotide binding sites, which interconvert, thus changing the afiity of the sites for the nucleotides. As the T site converts into a O site, ATP is released and another cycle starts. Direct observation of the rotation of the γ subunit of CF_1 has provided strong experimental support for the binding change mechanism.

Mode of Action of Some Herbicides

Herbicides of one major class—about half of the commercially important compounds—act by interrupting photosynthetic electron flow. A shows the chemical structure of two of these compounds. The precise sites of action of many of these agents have been found to lie either at the reducing side of photosystem I (for example, in paraquat) or in the quinone acceptor complex in the electron transport chain between the two photosystems.

Paraquat acts by intercepting electrons between the bound ferredoxin acceptors and NADP and then reducing oxygen to superoxide (O_2^-). Superoxide is a free radical that reacts nonspecifically with a wide range of molecules in the chloroplast, leading to the rapid loss of chloroplast activity. Lipid molecules in cell membranes are especially sensitive.

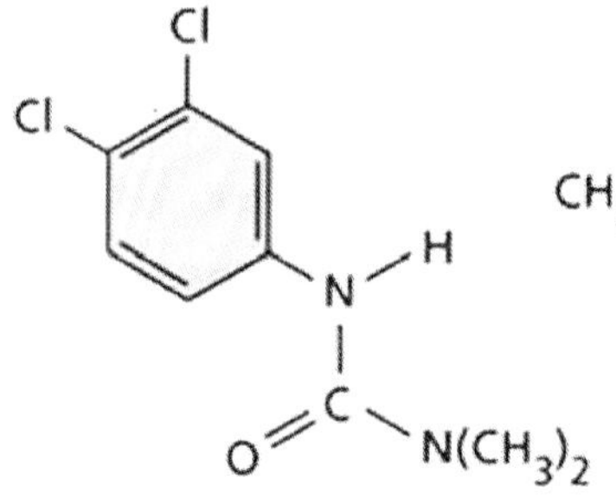

DCMU (diuron)
(Dichlorophenyl-
dimethylurea)

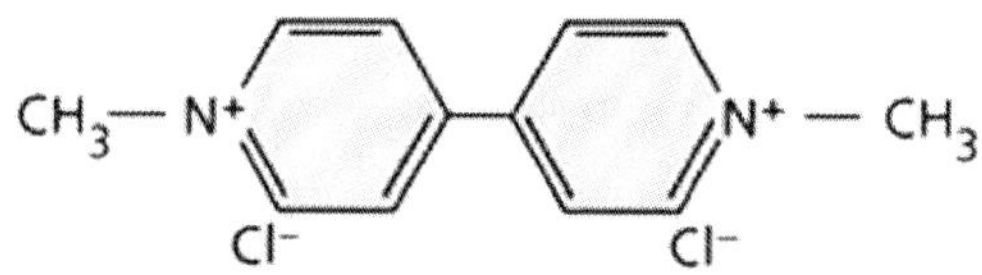

Paraquat
(methyl viologen)

Figure: *The use of herbicides to kill unwanted plants is widespread in modern agriculture. Many different classes of herbicides have been developed, and they act by blocking amino acid, carotenoid, or lipid biosynthesis or by disrupting cell division. Understanding the mode of action of herbicides has been an important tool in research on plant metabolism and has facilitated their application under different agricultural practices (Ashton and Crafts 1981). Here, the chemical structure of two herbicides that block photosynthetic electron flow are shown. DCMU is also known as diuron. Paraquat has acquired public notoriety because of its use on marijuana crops.*

The herbicides that act on the quinone acceptor complex compete with plastoquinone for the Q_B binding site. If herbicide is present, it displaces the oxidized form of plastoquinone and occupies the specific binding site for the quinone acceptor, which is thought to lie on the D1 herbicide-binding protein. The herbicide is not able to accept electrons, so the electron is unable to leave Q_A, the first quinone acceptor. Thus, the binding of herbicide effectively blocks electron flow and inhibits photosynthesis. Many herbicides that act in this manner also inhibit electron flow in photosynthetic bacteria that have quinone-type electron acceptor complexes.

In recent years, herbicide-resistant biotypes of common weeds have appeared in areas where a single type of herbicide has been used continuously for several years. These biotypes can be orders of magnitude more resistant to certain classes of herbicides than the nonresistant plant is. In several cases the resistance factor has been traced to a single amino acid substitution in the D1 protein.

This change, presumably in the quinone (and herbicide) binding region of the peptide, lowers the binding affinity of the herbicide, making it much less effective.

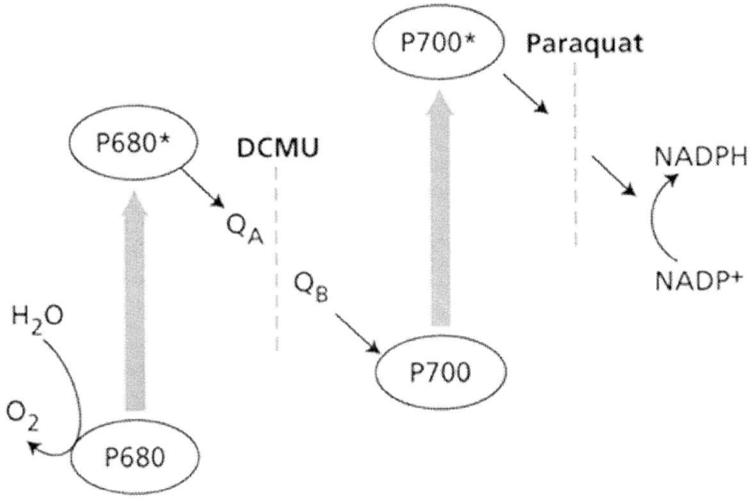

Figure: *Sites of action of herbicides. Many herbicides, such as DCMU, act by blocking electron flow at the quinone acceptors of photosystem II, by competing for the binding site of plastoquinone that is normally occupied by Q_B. Other herbicides, such as paraquat, act by accepting electrons from the early acceptors of photosystem I and then reacting with oxygen to form superoxide, O_2^-, a species that is very damaging to chloroplast components, especially lipids.*

The possibility of fine-tuning the herbicide sensitivity of crop plants by making subtle changes in the proteins of the PS-II reaction centre has created a great deal of interest in the agricultural chemical industry. Through biotechnology, it is now possible to make a crop plant that is resistant to a particular herbicide, which can then be applied to control undesirable plants that are not resistant. The success of this approach will depend on whether undesirable side effects of the herbicide-resistant mutations can be controlled and on how rapidly the weeds acquire resistance through natural selection or gene transfer. Thus, it is a race between science and evolutionary change.

A complementary approach is to insert resistance factors to certain pests, such as insects, into the plant via genetic engineering. The plants then produce the toxic species themselves, eliminating the need for insecticides.

Chlorophyll Biosynthesis

In the first phase of chlorophyll biosynthesis, the amino acid glutamic acid is converted to 5-aminolevulinic acid (ALA). This reaction is unusual in that it involves a covalent intermediate in which the glutamic acid is attached to a transfer RNA molecule. This is one of a very small number of examples in biochemistry in which a tRNA is utilized in a process other than protein synthesis. Two molecules of ALA are then condensed to form porphobilinogen (PBG), which ultimately form the pyrrole rings in chlorophyll. The next phase is the assembly of a porphyrin structure from four molecules of PBG. This phase consists of six distinct enzymatic steps, ending with the product protoporphyrin IX.

All the biosynthesis steps up to this point are the same for the synthesis of both chlorophyll and heme. But here the pathway branches, and the fate of the molecule depends on which metal is inserted into the centre of the porphyrin. If magnesium is inserted by an enzyme called magnesium chelatase, then the additional steps needed to convert the molecule into chlorophyll take place; if iron is inserted, the species ultimately becomes heme.

The next phase of the chlorophyll biosynthetic pathway is the formation of the fifth ring (ring E) by cyclization of one of the propionic acid side chains to form protochlorophyllide. The pathway involves the reduction of one of the double bonds in ring D, using NADPH. This process is driven by light in angiosperms and is carried out by an enzyme called protochlorophyllide oxidoreductase (POR). Non-oxygen-evolving photosynthetic bacteria carry out this reaction without light, using a completely different set of enzymes. Cyanobacteria, algae, lower plants, and gymnosperms contain both the light-dependent POR pathway and the light-independent pathway. Seedlings of angiosperms grown in complete darkness lack chlorophyll, because the POR enzyme requires light. These *etiolated* plants very rapidly turn green when exposed to light. The final step in the chlorophyll biosynthetic pathway is the attachment of the phytol tail, which is catalyzed by an enzyme called chlorophyll synthetase.

The elucidation of the biosynthetic pathways of chlorophylls and related pigments is a difficult task, in part because many of the enzymes are present in low abundance. Recently, genetic analysis has been used to clarify many aspects of these processes.

9

Biological Activity of Abscisic Acid

The features shown to be essential for biological activity include the carboxyl group, the tertiary hydroxyl group, the 2-*cis*, 4-*trans*-pentadienoic side chain, a 4'-ketone, and a double bond in the cyclohexane ring. Catabolic products of ABA that are present in the tissue and are missing any of these groups are biologically inactive.

The Bioassay of ABA

Coleoptile growth, the classic bioassay devised for auxins, is also used for ABA detection in plant extracts by measurement of coleoptile growth inhibition. This bioassay has adequate sensitivity (minimum detectable level is 10^{-7} *M*) and shows a linear response in the range of 10^{-7} to 10^{-5} *M*, but it has some disadvantages. Plant extracts have other promoters and inhibitors, which decrease the specificity of the bioassay and make preliminary purification steps necessary.

Other available bioassays for ABA include inhibition of germination, inhibition of gibberellic acid– induced α-amylase synthesis in cereal aleurone layers, induction of leaf abscission, and stomatal closure. Stomatal closure is highly specific for ABA because it is affected little by other plant growth regulators. Additional advantages of this bioassay include a fast response of guard cells to ABA, high sensitivity (minimum detectable level 10^{-9} *M*), and a linear response over a wide range of concentrations.

Proteins Required for Desiccation Tolerance

LEA protein sequences are highly conserved among a wide variety of species. They are related to two other groups of proteins, the RAB (*r*esponsive to *AB*A) and DHN (*de*h*y*dri*n*) proteins. These protein classes accumulate in a variety of tissues experiencing cellular dehydration, e.g. maturing embryos and pollen, and vegetative tissues exposed to drought stress, including extreme desiccation of "resurrection plants" (*Craterostigma plantagineum*).

The LEA, RAB, and DHN proteins are all water soluble, basic, rich in glycine and lysine, and low in hydrophobic residues. As a result they are extremely hydrophilic and stable to boiling. These features led Dure and colleagues (1989) to propose that they function specifically in the protection of membranes and proteins against desiccation damage, possibly by binding water tightly or providing hydrophilic interactions in the absence of free water, and by preventing crystallization of cellular components through their ability to act as stabilizing "solvents," similar to the effects of sugars. One of these proteins confers increased osmotolerance in transgenic yeast.

Types of Seed Dormancy and the Roles of Environmental Factors

During seed maturation, the embryo enters a quiescent phase in response to desiccation. Seed germination can be defined as the resumption of growth of the embryo of the mature seed; it depends on the same environmental conditions as vegetative growth does. Water and oxygen must be available, the temperature must be suitable, and there must be no inhibitory substances present.

However, in many cases a viable (living) seed will not germinate even though all the necessary environmental conditions for growth are satisfied. This phenomenon is termed seed dormancy. Seed dormancy introduces a temporal delay in the germination process that provides additional time for seed dispersal over greater geographical distances. It also maximizes seedling survival by preventing germination under unfavourable conditions.

Two types of seed dormancy have been recognized, coat-imposed dormancy and embryo dormancy. Coat-imposed dormancy is dormancy imposed on the embryo by the seed coat and other enclosing tissues, such as endosperm, pericarp, or extrafloral organs. The embryos of such seeds will germinate readily in the presence of water and oxygen once the seed coat and other surrounding tissues are either removed or damaged. There are five basic mechanisms of coat-imposed dormancy:

Prevention of water uptake. Prevention of water uptake by the seed coat is a common cause of seed dormancy in families found in arid and semiarid regions, especially among legumes, such as clover (*Trifolium* spp.) and alfalfa (*Medicago* spp.). Waxy cuticles, suberized layers, and lignified sclereids all combine to restrict the penetration of water into the seed.

Mechanical constraint. The first visible sign of germination is typically the radicle breaking through the seed coat. In some cases, however, the seed coat may be too rigid for the radicle to penetrate. Nuts with hard, lignified shells are examples of dormancy caused by mechanical constraint. Such shells must be broken by biotic or environmental forces for the seed to germinate. Even nonlignified tissues, such as the endosperm of lettuce seeds, can suppress expansion of the embryo. For the seeds to germinate, the endosperm cell walls must be weakened by the production of cell wall–degrading enzymes.

Interference with gas exchange. Some seed coats are considerably less permeable to oxygen than an equivalent thickness of water is—e.g., less by a factor of about 10^4 in seeds of cocklebur (*Xanthium pennsylvanicum*). This lowered permeability to oxygen suggests that the seed coat inhibits germination by limiting the oxygen supply to the embryo. In support of this idea, investigators can break the dormancy of such seeds either by making a small hole in the coat with a pin (without weakening the coat mechanically), or by treating the coat with concentrated oxygen. However, other studies suggest that the oxygen consumption of the embryos from such seeds is considerably less than the amount of oxygen able to penetrate the seed coats under normal aerobic conditions. Thus the role of oxygen impermeability in seed coat dormancy remains unresolved.

Retention of inhibitors. The seed coat may prevent the escape of inhibitors from the seed. For example, growth inhibitors readily diffuse out of isolated *Xanthium* embryos, but not from intact seeds.

Inhibitor production. Seed coats and pericarps may contain relatively high concentrations of growth inhibitors that can suppress germination of the embryo. ABA is a common germination inhibitor present in these maternal tissues. In certain cases where repeated washing (leaching) removes dormancy, the effect is thought to be due to the loss of such inhibitory compounds.

The second type of seed dormancy is embryo dormancy, which refers to a dormancy that is inherent in the embryo and is not due to any influence of the seed coat or other surrounding tissues. In some cases, embryo dormancy can be relieved by amputation of the cotyledons. Species in which the cotyledons exert an inhibitory effect include European hazel (*Corylus avellana*) and European ash (*Fraxinus excelsior*). A fascinating demonstration of the cotyledon's ability to inhibit growth is found in species (e.g., peach) in which the isolated dormant embryos germinate but grow extremely slowly to form a dwarf plant. If the cotyledons are removed at an early stage of development, however, the plant abruptly shifts to normal growth.

Embryo dormancy is thought to be due to the presence of inhibitors, especially ABA, as well as the absence of growth promoters, such as GA (gibberellic acid). The loss of embryo dormancy is often associated with a sharp drop in the ratio of ABA to GA.

Seed Dormancy May be Primary or Secondary

Different types of seed dormancy also can be distinguished on the basis of the timing of dormancy onset rather than the cause of the dormancy.

- Seeds that are released from the plant in a dormant state are said to exhibit primary dormancy.
- Seeds that are released from the plant in a nondormant state but which become dormant if the conditions for germination are unfavourable exhibit secondary dormancy. For example, seeds of *Avena sativa* (oat) can become dormant in the presence of temperatures higher than the maximum for germination, whereas seeds of *Phacelia dubia* (small-flower scorpionweed) become dormant at temperatures below the minimum for germination. The mechanisms of secondary dormancy are poorly understood.

Environmental Factors Control the Release from Seed Dormancy

Various external factors release the seed from dormancy, and dormant seeds typically respond to more than one condition. Many seeds lose their dormancy when their moisture content is reduced to a certain level by drying. This method of breaking seed dormancy is called afterripening, and is usually performed in a special drying oven. On the other hand, if the seed becomes too dry (5% water content or less), the effectiveness of afterripening is diminished.

Another factor that can release seeds from dormancy is low temperature, or chilling. Many seeds require a period of cold (0 to 10°C) while in a fully hydrated (imbibed) state in order to germinate. In temperate-zone species, this requirement is of obvious survival value, since such seeds will germinate not in the fall, but only in the following spring. Chilling seeds to break their dormancy is a time-honored practice in horticulture and forestry and traditionally has been referred to as *stratification.* This term is derived from the old agricultural practice of allowing seeds with a chilling requirement to overwinter outdoors in layered mounds of moist soil. Today the seeds are simply stored in a refrigerator.

The third external factor that plays an important role in breaking seed dormancy is light. Many seeds have a light requirement for germination, which may involve only a brief exposure, as in the case of lettuce, an intermittent treatment (e.g., succulents of the genus *Kalanchoe*), or even a specific photoperiod involving short or long days. The main sensor for light-regulated seed germination. Interestingly, all light-requiring seeds exhibit seed coat dormancy, and removal of the outer tissues of the seed allows the embryo to germinate in the absence of light. The effect that light has on the embryo is thus to enable the radicle to penetrate the seed coat. This penetration often involves some enzymatic weakening of the enclosing tissues.

The Longevity of Seeds

How long can a seed remain dormant and still remain viable? Seed longevity is of practical importance because of ongoing efforts to preserve plant genetic resources for future agricultural crops by setting up seed gene banks. Spectacular claims of dormant seeds that remained viable for thousands of years have been made, but are considered highly controversial. Extreme examples of ancient seeds that have been reported to retain their viability include submerged lotus seeds found in a 3,000-year-old boat near Tokyo, barley seeds from the 3,000-year-old tomb of King Tutankhamen, and arctic lupine seeds associated with rodent burrows determined to be 14,000 years old. In most cases, however, a scientifically rigorous examination of the data has either disproved or raised serious doubts about the claimed antiquity of the seeds (Bewley and Black 1994).

The most serious criticism is that the dates of the seeds have been inferred from the age of their immediate surroundings or associated artifacts rather than from direct measurements of the seeds themselves. This question raises the possibility that the seeds are much younger than the site at which they were found. In fact, microscopic examination of authenticated ancient cereal grains from various Middle Eastern sites has shown that cells and tissues of these seeds are no longer intact. Such seeds are clearly inviable.

Although the most sensational claims for seed longevity are almost certainly bogus, seeds of *Canna compacta* apparently can live for at least 600 years. Viable *Canna* seeds were obtained from inside a walnut in a tomb in Argentina. The *Canna* seeds had apparently been inserted into the immature seeds of a growing walnut fruit before the hard outer shell formed. Once the shell hardened and the nut dried out, the result was a rattle. Native people strung the rattles together to form a necklace. In this case, the seeds had to be at least as old as the walnut shell, and carbon dating of the shell indicated that it was about 600 years old.

Plant collections begun during the late eighteenth century in Europe—for example, at the British Museum in London and at the Museum of Natural History in Paris—have been reliable sources of seeds for viability determinations. *Cassia multijuga* (false sicklepod) seeds, collected in 1776, were still viable after being tested in 1934. In 1879, W. J. Beal initiated the longest-running experiment on seed longevity by burying the seeds of 21 different species in unstoppered bottles in a sandy hilltop near the Michigan Agricultural College in East Lansing. After 100 years, only one species, *Verbascum blattaria* (moth mullein), remained viable. Why are some seeds able to survive in a dormant state for hundreds of years, whereas others lose viability in less than 5 years? Almost nothing is known about the mechanisms that determine the longevity of seeds. If these mechanisms could be understood, we might someday be able to greatly increase the seed longevity of agriculturally important species and varieties, thereby enhancing our ability to preserve plant genetic resources for generations to come.

Genetic Mapping Of Dormancy: Quantitative Trait Locus (QTL) Scoring of Vegetative Dormancy Combined with a Candidate Gene Approach

Complex traits, e.g. dormancy, are affected by many different, often unlinked genes. Consequently, attempts to map them result in continuous, or quantitative, variation within a mapping population.

Quantitative trait locus (QTL) mapping is used in these instances to map individual genes affecting the trait. In QTL mapping, inbred lines that differ in the trait of interest are crossed and an F_2 population is produced. The F_2 population has continuous variation for the trait of interest, and the goal of the analysis is to determine which chromosomal segments contain genes affecting the trait. Using marker loci along each chromosome (loci that have different alleles between the parental inbred lines), the genotypes of entire chromosomal segments in each F_2 individual can be inferred as being homozygous for one or the other parental lines or heterozygous. If the inheritance of a chromosomal segment is correlated with the trait of interest, then a gene affecting the trait is likely to be between the two marker loci. By analysing chromosomal segments along each of the chromosomes, the number and locations of genes affecting the quantitative trait can be determined.

Candidate genes are specific loci, identified by predicted biochemical function or homology to regulatory genes from other species, that are suspected of affecting the trait of interest. Molecular differences between the parental alleles at these loci can be used to determine whether they map to the same region as any of the QTLs, thereby supporting the hypothesis that a given gene contributes to regulation of the trait.

ABA-Induced Senescence and Ethylene

The ethylene has also been found to play a role in the senescence of oat leaf segments, but ABA appears to be the initiating agent, whereas ethylene appears to exert its effects at a later stage. To determine whether the ABA effects are mediated by ethylene, the effects of ABA and ethylene on senescence of leaf discs from wild-type and ethylene-insensitive mutants were compared. Chlorophyll loss from untreated leaf discs of the ethylene-insensitive mutant was much slower than that from the wild type, in the light and in the dark.

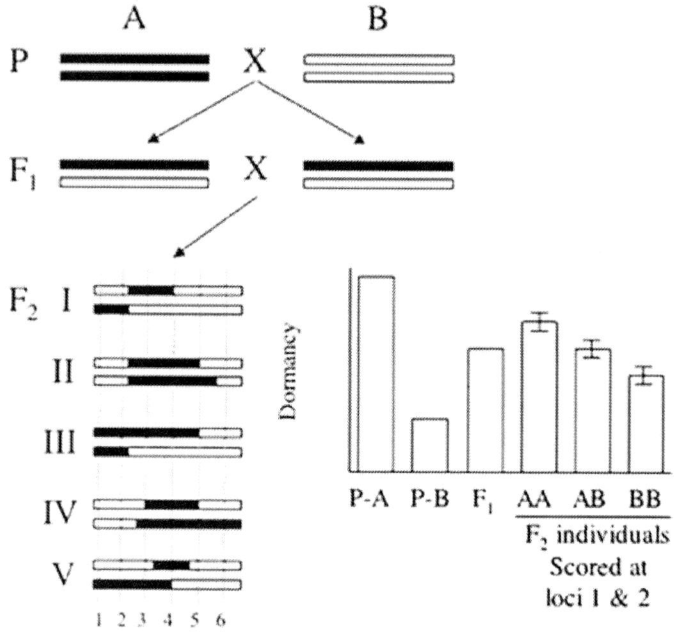

Figure: *Two parental (P) inbred lines, A & B, which differ for a trait of interest (e.g. dormancy) are crossed to produce heterozygote F_1 progeny which have an intermediate degree of dormancy. Only one chromosome pair is depicted for simplicity. The F_1 progeny are crossed to produce a large segregating F_2 population (usually 100-500 individuals are analysed, here only five individuals, I-V, are depicted). Because of recombination during F_1 meiosis, the F_2 individuals have varying genetic combinations of the original inbred lines (depicted by shading of chromosomes) and have continuous variation, from high to low, in degree of dormancy. In order to estimate whether segments of the chromosome are homozygous for one of the parental-lines or heterozygote, the F_2 individuals are scored for marker loci (1-6) along the chromosome. For example, loci 1 & 2 are identified as homozygous for A-line alleles in individual III, as heterozygous in individuals I & V and as homozygous for B-line alleles in individuals II & IV. The F_2 individuals are also scored for their dormancy. After scoring many F_2 individuals a correlation between the genotypes at locus 1 & 2 and phenotype is determined and therefore this segment likely contains at least one gene affecting dormancy (bar graph). The analysis is repeated for each chromosomal segment to identify which ones are likely to contain genes affecting dormancy.*

Ethylene accelerated the chlorophyll loss from the wild-type leaf discs, but it had no effect on the yellowing of the ethylene-insensitive mutant discs. In contrast, treatment with ABA stimulated chlorophyll loss in *both* the wild-type and the mutant discs, whether they were held in the light or in the dark (Zacarias and Reid 1990). These results indicate that the promotion of senescence by ABA is not mediated through its stimulation of ethylene production.

Yellow Cameleon: A Noninvasive Tool for Measuring Intracellular Calcium

Cytoplasmic free calcium ($[Ca^{2+}]_{cyt}$) acts as a secondary messenger in a variety of signalling pathways and it appears that the frequency, magnitude and localization of $[Ca^{2+}]_{cyt}$ oscillations are important determinants of specificity. "Traditional" techniques for measuring Ca^{2+} have made use of microinjected Ca^{2+} sensitive indicator dyes, but these methods are technically difficult and invasive. An alternative strategy makes use of Ca^{2+} binding proteins that can be transgenically expressed throughout an organism, eliminating the need for microinjection.

The Ca^{2+} indicator yellow cameleon is a fusion protein composed of four domains: two modified green fluorescent proteins (GFPs) linked via calmodulin (CaM) and a CaM-binding peptide ("M13").

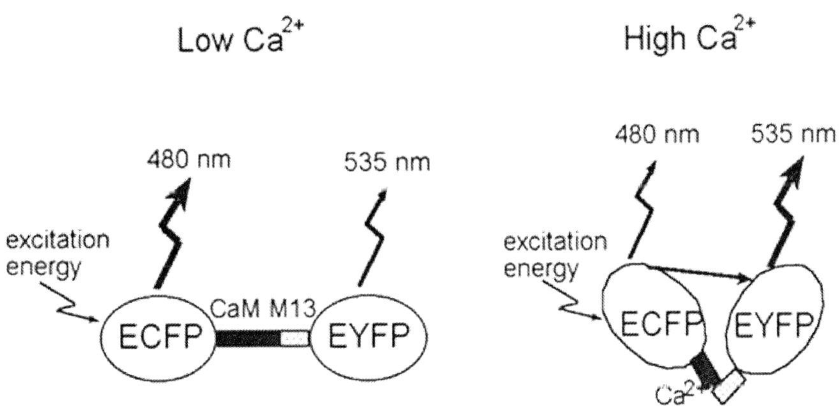

***Figure:** The binding of calcium brings the two ends of the yellow cameleon protein together, altering the ratio of 535 nm to 480 nm emission.*

The GFPs are modified to shift their emission wavelengths toward blue (enhanced cyan fluorescent protein, ECFP) or yellow (enhanced yellow fluorescent protein, EYFP). When the CaM domain binds to Ca^{2+}_{cyt} it associates with the CaM-binding peptide, producing a conformational change that brings the modified GFP domains closer together.

Following exposure to a laser, excitation energy from ECFP (the higher-energy fluorescent protein) is passed to EYFP resulting in increased emission by this lower-energy fluorescent protein. Consequently, increased $[Ca^{2+}]_{cyt}$ results in increased emission at 535 nm relative to that at 480 nm, and the 535/480 emission ratio is proportional to the $[Ca^{2+}]_{cyt}$ (Miyawaki et al 1997; Allen et al. 1999).

The Two-Hybrid System

Many cellular proteins interact with a variety of other proteins to perform their physiological functions. Identification of interacting proteins can be accomplished by biochemical assays, such as co-purification of a complex, but these require relatively stable interactions and the ability to identify components of the complex immunologically or based on amino acid sequence and composition. Alternatively, the two-hybrid assay is a genetic approach that takes advantage of the modular nature of proteins to identify interacting domains.

Although there are several variations, in all cases two fusion proteins are co-expressed in a cell line that will activate expression of a reporter gene if the fusion proteins interact.

For example, the yeast transcription factor GAL4 can be separated into distinct DNA binding (BD) and transcription activation (AD) domains. If hybrid genes are constructed such that they encode fusion proteins BD-protein X and AD-protein Y, interactions between proteins X and Y will bring the GAL4 domains close enough to reconstitute a functional transcription factor that can trans-activate a GAL4-responsive promoter.

The cell lines often contain several GAL4-responsive promoter-reporter fusions, whose activity can be scored either qualitatively or quantitatively. Qualitative scoring (+/- growth) makes it possible to screen a cDNA library for fusions encoding interacting proteins by requiring expression of a gene whose activity can complement an auxotrophy. Quantitative scoring (e.g. b-galactosidase activity from a *lacZ* reporter) permits comparisons of the relative strength of interactions among a variety of protein pairs.

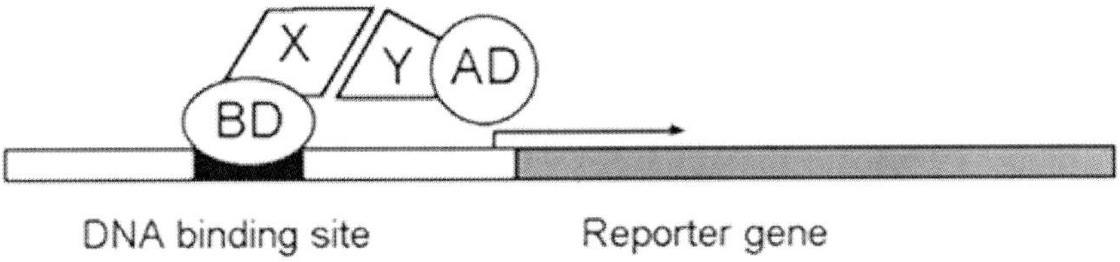

Figure: *Two fusion proteins containing the GAL4 DNA binding (BD) and transcription activation (AD) domains can reconstitute a functional transcription factor that can trans-activate a GAL4-responsive promoter if the interactions between the two proteins, X and Y, will bring the GAL4 domains close enough together.*

The Control of Flowering

Regulation of Juvenility by the TEOPOD (TP) *Genes in Maize*

The regulation of the transition from juvenile to adult (i.e., the attainment of competence) is influenced by a wide variety of conditions that affect the growth rate, and by transmissible factors such as carbohydrates and gibberellins.

Hence it has been difficult to determine exactly what regulates the phase change during normal development. The analysis of mutations that affect the timing of phase changes, called heterochronic mutations, provides the opportunity to identify specific genes that regulate juvenility. The best-studied example is in maize.

The juvenile and adult phases of corn plants are characterized by differences in vegetative morphology. Scott Poethig and his colleagues at the University of Pennsylvania have been characterizing the heterochronic *teopod* (*tp*) mutants of maize: *tp1*, *tp2* and *tp3*.

These semidominant mutations cause cells and tissues that would normally develop as adult structures to become more juvenile. In *tp* plants, leaves in the normally adult part of the shoot are morphologically intermediate between juvenile and adult leaves. Whereas in the wild type, tassels bear male flowers and ears bear female flowers, in *tp* mutants both tassels and ears possess leaves in addition to flowers.

However, *tp* mutations affect neither the rate nor the duration of growth of the plant, nor do they interfere with the timing of tassel formation. The *teopod* mutations thus specifically affect the transition from juvenile to vegetative adult phase without affecting the transition to the reproductive phase.

Figure: *Phenotypes of three* teopod *mutants and the wild-type maize plant. From left to right:* tp1, tp2, tp3, *and wild type.*

The possible interaction between the *TEOPOD* genes and the gibberellins has been investigated. In GA-deficient dwarf mutants of maize, the transitions from juvenile to adult vegetative development and from adult vegetative to reproductive development occur later than in wild-type plants.

Thus, endogenous GAs are required for the normal timing of phase changes.

Flowering of Juvenile Meristems Grafted to Adult Plants

By definition, juvenile plants cannot be induced to flower with an inductive photoperiod, suggesting either that the shoot apical meristems of juvenile plants are not competent to flower, or that juvenile leaves produce insufficient amounts of a floral stimulus. The LSDP *Bryophyllum* passes through a juvenile phase, which lasts from 5 months to a year (depending on the species), during which it cannot be induced to flower by transfer from LD to SD.

However, when juvenile *Bryophyllum* shoots were grafted onto flowering plants, the juvenile shoots flowered promptly and profusely. This result indicates that the shoot apical meristems of juvenile plants are competent to flower, but fail to do so because the juvenile leaves produce insufficient amounts of the floral stimulus.

Age is another factor that can affect production of the floral stimulus. Grafting studies with the SDP *Perilla* have shown that the leaves at the top of a plant are stronger donors of the floral stimulus than the leaves near the base of a plant, in keeping with the juvenile character of the basal leaves.

The experiment was performed as follows. Two groups of *Perilla* plants were sown at different times. When the first group had produced five pairs of leaves, the second group had only produced two. The fifth and the second pairs of leaves from the two sets of plants were grafted onto stocks and maintained under long days.

Once the graft unions were established, the grafted leaves were induced with short days by daily enclosing them in opaque bags. Leaves from the fifth node needed only 14 short days to induce flowering while leaves from the second node required 28 short days. This finding shows that the more juvenile leaves from the second node produce less floral stimulus, and thus require more inductive photoperiods to induce flowering, than the fully adult leaves higher up the shoot.

Support for the Role of Blue-Light Regulation of Circadian Rhythms

The role of blue light in regulating both circadian rhythmicity and flowering is also supported by studies with an *Arabidopsis* flowering-time mutant: elf3 (early flowering 3). The *elf3* mutant flowers earlier than the wild type under a variety of growth conditions and is insensitive to photoperiod; that is, it is day-neutral.

Transformation of *elf3* with the luciferase reporter construct revealed a defect in circadian rhythmicity. The mutant displayed no circadian rhythmicity when grown in continuous light. Significantly, the *elf3* mutant also has a long hypocotyl when grown in constant white light, similar to the *hy* mutants, and is less responsive than the wild type to all wavelengths of light, but especially to blue light.

This finding suggests that the circadian clock is intact in the mutant, but that the transduction of light signals to the clock is impaired.

The fact that the *elf3* mutant is more impaired in its response to blue light than in its response to red light further suggests that a blue-light photoreceptor plays a role in the photoperiodic control of flowering.

Genes That Control Flowering Time

As we have seen, the Maryland Mammoth mutation converts a day-neutral variety of tobacco into an SDP. Analysis of this flowering-time mutant by W. Garner and H. Allard led to the discovery that plants can measure photoperiod as an environmental cue for floral induction, and it identified the Maryland Mammoth gene as one of the genes that regulates this process.

The most extensive studies of flowering-time genes have been performed in pea (*Pisum sativum*) and *Arabidopsis thaliana*. In these species a range of mutations can alter various aspects of the photoperiodic control of flowering. For example, mutations in certain genes convert these LDPs into day-neutral plants that can flower rapidly.

Mutations in other genes result in day-neutral plants that exhibit a long delay in flowering. Because these mutations eliminate the response to photoperiod, the normal (wild-type) products of these genes are thought to be involved in photoperiodism. Other mutations can delay or promote flowering without affecting the photoperiodic response. These mutants are thought to affect genes involved in the *autonomous* regulation of flowering.

Since most biological regulatory systems are under both positive and negative control, it is not surprising to find mutations that promote or delay flowering. Inasmuch as most mutations result in a loss or reduction of gene activity, the wild-type role of genes that delay flowering when mutated is to promote flowering.

Similarly, mutations that promote flowering are likely to occur in genes that normally inhibit the process. These genes are promoters and inhibitors of flowering in genetic terms, but the mutant phenotype alone does not usually reveal the flowering mechanism that is affected. For example, a mutation that eliminates the response to photoperiod could either block the production of a flowering hormone (floral stimulus) or, alternatively, interfere with the ability of the target cells in the meristem to receive the hormonal signal (meristem competence).

Arabidopsis plants containing mutations that promote or delay flowering have been isolated. The wild type and a *gigantea* (*gi*) mutant that were grown in inductive photoperiods (long days). The plants are shown at the stage when flower buds have become visible.

Flowering in the *gigantea* mutant is delayed; thus this plant has been growing approximately three times as long as the wild type and has formed many more leaves from the primary apical meristem. Because a mutation in the *GIGANTEA* gene delays flowering, the normal role of this gene is to promote flowering. The wild type and an *elf3* mutant plant that were grown in noninductive photoperiods (short days). The mutant phenotype indicates that the normal role of the *ELF3* gene is to repress flowering.

One way to explore the role of these genes is through grafting studies. For example, if the apical bud from a mutant plant with a delayed-flowering phenotype flowers normally when grafted onto a wild-type shoot, the mutant shoot apex is clearly competent to receive and respond to flowering signals. One might then suspect that the mutation affects the production of flowering signals by leaves.

If this were the case, a wild-type shoot apex region should exhibit delayed flowering when grafted onto a mutant shoot. Grafting studies of this sort have identified two types of genes in pea: those that control the production of transmissible promoters and inhibitors of flowering, and those that control the sensitivity of the shoot apical meristem to these signals.

For example, the *GIGAS* gene is thought to be involved in the production of a transmissible flowering promoter. The wild-type pea exhibits a quantitative LDP phenotype. Plants homozygous for the recessive *gigas* mutation exhibit severely delayed flowering. For example, when grown under a 16-hour photoperiod, the wild type flowers after producing 16 leaves, whereas the *gigas* mutant flowers after producing 43 leaves. Delayed flowering in *gigas* plants could be caused either by the presence of inhibitors or by the absence of a floral stimulus.

To test for inhibitors, researchers grafted the shoot of an early flowering day-neutral line onto *gigas* stock. The time of flowering of the scion was unaffected by grafting to the mutant, indicating that the mutant does not produce flowering inhibitors.

However, when a *gigas* shoot was grafted onto either the early-flowering or the wild-type stock, flowering was greatly accelerated. Instead of flowering after producing 40 leaves or more, the plants flowered after producing only 14 or 23 leaves, respectively.

Because the *gigas* mutant shoot apex flowers readily after grafting, its ability to respond to the flowering stimulus from the leaves is not affected by the mutation. The mutation is therefore thought to affect the production of a floral stimulus.

Another approach to elucidating the floral induction process is to isolate the genes identified by mutations and to study their properties. Once the gene is cloned, the encoded protein can be overexpressed in *E. coli* and purified in sufficient quantities to study its biochemical properties and to produce specific antibodies. The antibodies can be used to determine the tissues in which the protein is produced, as well as its location within the cell.

The localization may provide clues to the function of the protein. For example, a protein in the shoot apex may be involved in the reception of flowering signals, whereas a protein present in the leaves may be involved in the production of flowering signals.

The *constans* mutants of *Arabidopsis* are insensitive to day length and flower later than the wild type under long-day conditions. The *CONSTANS* gene of *Arabidopsis* has been cloned and shown to encode a protein that shows similarities to zinc finger transcription factors, suggesting that it acts as a transcriptional regulator. The wild-type *CONSTANS* gene is expressed in leaves, and the level of *CONSTANS* mRNA increases during inductive photoperiods.

This pattern of expression, along with the photoperiod-insensitive mutant phenotype, implicate this gene in the photoperiodic control of flowering.

The Different Effects of Two Different Gibberellins on Flowering (Spike Length) and Elongation (Stem Length)

In the long-day plant *Lolium temulentum*, GA_{32} strongly promotes flowering, but has only a small effect on stem elongation. In contrast, GA_1 strongly promotes stem growth but has only a small effect on flowering. Gibberellin-treated plants were maintained in short days and, for comparison, the response to a single 24-hour day is shown. This long day strongly promotes flowering but has almost no effect on stem length.

The Influence of Cytokinins and Polyamines on Flowering

Sometimes changes normally associated with flowering can be caused by treatments that do not themselves bring about floral initiation. For example, one of the earliest events observed at the shoot apex following photoperiodic induction is a transitory increase in the number of cells undergoing mitosis.

When cytokinins are applied to the stem apex of the long-day plant white mustard (*Sinapis alba*), they cause an increase in mitotic activity similar to that caused by exposure to a single long day. However, whereas the long day induces flowering, the cytokinin treatment does not. Cytokinin may therefore be a component of the floral stimulus, although it is insufficient by itself to bring about flowering.

Other potential components include the polyamines. For example, photoperiodic induction of the LDP *Sinapis alba* was correlated with a large increase in the concentration of putrescine, the major polyamine, in the phloem exudate from the leaves. Spraying the leaves with an inhibitor of putrescine biosynthesis lowered the level of putrescine in the phloem exudate and also inhibited flowering. This result suggests that putrescine may be a component of the floral stimulus in *S. alba*.

In addition to receiving signals from the leaves, the shoot apex may accumulate flowering signals from the roots. Classic experiments with cuttings have shown that the initiation and growth of adventitious roots inhibit flowering in the shoot. In many species, root removal promotes flowering. Thus, at least in certain species, the root appears to produce one or more flowering inhibitors. In other cases, the roots may contribute floral activators.

For example, both GA and cytokinin are synthesized in the root and transported to the shoot via the xylem. During photoperiodic induction in *Sinapis alba,* the amount of cytokinin present in the xylem exudate from the root increases considerably. This finding indicates that signals emanating from the leaf are transported both to the shoot apex and to the root during photoperiodic induction. In response to the signal from the leaves, the roots produce and export additional chemical signals that modulate the flowering response.

The Contrasting Effects of Phytochromes A and B on Flowering

Phytochrome is encoded by members of a multigene family, and it is possible that the effects of red and far-red light described in the previous section are mediated by different types of phytochrome (A and B). In LDPs such as *Arabidopsis,* pea (*Pisum sativum*), and sorghum (*Sorghum bicolor*), phytochrome B (phyB) appears to be an *inhibitor* of flowering, since mutations in the gene *PHYB* that eliminate or reduce the amount of the phyB protein cause more rapid flowering.

The promotion of flowering by far-red light may result from a decrease in the amount of the Pfr form of phyB, which would have the same effect on the phyB Pfr levels as the mutations that reduce the total amount of phyB. As already discussed, red-light interruptions of a dark period prevent flowering in SDPs; therefore, the Pfr form of phyB may be an *inhibitor* of flowering in SDPs as well.

In the LDPs *Arabidopsis* and pea, mutations in the *PHYA* gene interfere with the promotion of flowering by long days; thus the Pfr form of phyA may promote flowering in LDPs. The effect of the *phyA* mutation is quite strong in pea (the mutant plants are essentially unable to respond to inductive photoperiods) whereas the *phyA* mutation in *Arabidopsis* has only a slight effect on the ability to respond to inductive photoperiods. The species difference in the effect of *phyA* mutations is probably due to the fact that pea relies entirely on phyA to sense inductive photoperiods, whereas *Arabidopsis* also uses a blue light photoreceptor for this purpose.

A Gene that Regulates the Floral Stimulus in Maize

Maize plants flower after producing a fixed number of leaves regardless of photoperiod—a characteristic of autonomous regulation. If the shoot apical meristem from a maize plant is excised, and the "explant" is cultured in vitro, it is possible to regenerate a complete plant that will eventually flower. As long as the explant includes 4-6 leaf primordia, the meristem "remembers" its developmental stage at the time of excision and produces only the additional number of leaves needed to make up the fixed amount before flowering.

However, if the explant contains fewer than 4 leaf primordia, the apex "forgets" its developmental stage and produces a complete set of new leaves before flowering.

The *indeterminate 1* (*id1*) mutant of maize is unable to undergo a normal transition to flowering, and produces many more leaves than the wild type. Moreover, the flowers produced by *id1* mutants are abnormal, having vegetative characteristics.

The *ID1* gene encodes a zinc-finger protein, suggesting that it functions as a transcriptional regulator. The expression pattern of *ID1* differs from that of many other flowering time control genes isolated from *Arabidopsis* in that it is expressed in young leaves rather than in the shoot apical meristem.

Are the effects of the *ID1* gene "cell autonomous", that is, restricted to those cells in which the gene is expressed, or does the *ID1* gene regulate the production of a transmissible signal that can affect cells beyond the site of gene expression? Traditionally, grafting experiments between the mutant and the wild type have been carried out to answer such questions, but maize (a monocot) lacks a vascular cambium and is therefore not suitable for grafting experiments.

In a second set of experiments, Colasanti et al. (1998) did the next best thing to grafting by analysing plants that were genetic chimeras (that is, containing cells with different genotypes) for *ID1* function. The mutant allele of *id1* analysed was caused by the insertion of a transposable element, and the spontaneous excision of this element during development generated clonal sectors of wild-type *ID1*-expressing cells against an overall mutant *id1* background. It was found that chimeric *id1* plants containing sectors of restored *ID1* function flowered earlier than completely mutant *id1* plants, even though the shoot apical meristems of these chimeric plants remained mutant. This experiment is comparable to grafting a wild type scion onto an *id1* mutant stock, and finding that the *id1* stock flowered earlier as a result. This observation in combination with the expression data provides compelling evidence for the involvement of *idt1* in the production in leaves of a transmissable signal from immature leaves that induces flowering in the apex.

The *Arabidopsis* genome contains several potential homologs of the maize *ID1* gene, but no *id1*-like mutants have been identified with a phenotype similar to that in maize. All of the genes involved in the autonomous pathway of *Arabidopsis* thus far seem to be expressed exclusively in the apical meristem.

Bibliography

Ashworth S.: *Seed to Seed,* Decorah, Seed Savers Publications, 1991.

Bag, A. K.: *Mathematics in Ancient and Medieval India,* Varanasi: Chaukhambha Orientalia, 1979.

Baruah, Akhil: *Advanced Morphology of Angiosperms,* Aavishkar, Delhi, 2008.

Berger, P.L.: *Pyramids of Sacrifice: Political Ethics and Social Change,* New York, Basic Books, 1974.

Bosworth, B.: *Capital Formation and Economic Policy.* Brookings Papers on Economic Activity, 1982

Boyer, J.S.: *Measuring the Water Status of Plants and Soils,* Academic Press, N.Y., 1995.

Brainerd, Charles J.: *The Origins of the Number Concept,* New York, Praeger, 1979.

Chadha, K. L. and Pareek O. P.: *Advances in Horticulture: Fruit Crops,* New Delhi, Malhotra Publishing House, 1993.

Chambers, R.: *Rural Development: Putting the Last First,* London, Longman, 1983.

Clawson, Calvin: *The Mathematical Traveller,* New York, Plenum Press, 1994.

Collymore L.: *Fruit Production in Barbados,* Port of Spain, Trinidad and Tobago, 1996.

Currah L. and Proctor F. J.: *Onions in Tropical Regions,* Kent, Natural Resources Institute, 1990.

Dabholkar, A.R.: *General Plant Breeding,* Concept, Delhi, 2006.

David White: *The Physiology and Biochemistry of Prokaryotes,* Oxford University Press, Delhi, 2007.

Ferentinos L.: *Proceeding of the Sustainable Taro Culture for the Pacific Conference*, Honolulu, HITAHR, 1993.

Fortuner, R.: *Nematode Identification and Expert System Technology*, New York, Plenum Press, 1988.

Georges, S.: *The Debt Boomerang: How Third World Debt Harms Us All*, Boulder, Westview Press, 1992.

Hardy B.: *Biology and Agronomy of Forage Arachis*, Cali, International Centre for Tropical Agriculture, 1994.

Harsh Vardhan Bhaskar: *Animal Physiology*, Campus Books, Delhi, 2008.

Hartley W.: *A Checklist of Economic Plants in Australia*, Melbourne, C.S.I.R.O., 1979.

Herminie Broedel Kitchen: *Soils and Crops : Diagnostic Techniques*, Satish Serial Publishing, Allahabad, 2004.

Hessayon D. G. Dr.: *The Vegetable Expert*, England, PBI, Publications, 1985.

Jeffers P.: *Evaluation of Four Onion Varieties in Montserrat*, Plymouth, CARDI, 1992.

Jones, R. M.: *Plant Resources of South-East Asia*, Wageningen, Pudoc Scientific Publishers, 1992.

Julia F.: *Fruits of Warm Climates*, Miami, Julia F. Morton Publisher, 1987.

Keshav Prasad Yadav: *Application of Morphometry in Geomorphology*, Radha Pub, Delhi, 2008.

Khanna, D.R. and P.R. Yadav: *Biology of Fishes*, Discovery, Delhi, 2004.

Kim J. S.: *Sources of Resistance to Viruses of Pepper (Capsicum spp.): a Catalogue*, Taipei, AVRDC, 1994.

Kroll R.: *Cut Flowers*, Wageningen, CTA, 1995.

Kunelius T.: *Annual Ryegrasses in Atlantic Canada*, Ottawa, Agriculture Canada, 1991.

Macself, A.J.: *Soils and Fertilizers*, Satish Serial Pub, Delhi, 2005.

Malins, A.: *Postharvest Handling of Pineapple and Mango*, Port of Spain, Trinidad and Tobago, 1992.

Menninger, Karl: *Number-Words and Number-Symbols*. London, MIT, 1969.

Mondal, M.S.: *Pollen Morphology and Systematic Relationship of the Family Polygonaceae*, BSI, Delhi, 1997.

Nobel, P. S.: *Physicochemical and Environmental Plant Physiology*, Academic Press, San Diego, 1999.

Oldham P.: *Cost of Production of Major Tree Crops in Dominica*, Roseau, Ministry of Agriculture, 1991.

Parry M.L.: *Climatic Change, Agriculture and Settlements*, Dawson Folkestone UK, 1978.

Pemberton, R. W.: *Predictable Risk to Native Plants in Weed Biological Control*, Oecologia, 2000.

Politycka, B. and Goswami, C. L.: *Plant Physiology: Research Methods*, Scientific, Delhi, 2007.

Qystein V. Sjaastad: *Physiology of Domestic Animals*, International Book Distributing Co., Delhi, 2005.

Ragone D.: *Breadfruit: Artocarpus Altilis (Parkinson) Fosberg*, Rome, International Plant Genetic Resources Institute, 1997.

Rao, S. Balaachandra: *Indian Mathematics and Astronomy*, Bangalore, Jnana Deep Publications, 1994.

Renuka Desai: *Plant Physiology*, Adhyayan, Delhi, 2008.

Robinson, D.H.: *Entomology : Principles and Practices*, Agrobios, Delhi, 2001.

Sanchez, P. A.: *Properties and Management of Soils in the Tropics*, John Wiley and Sons, New York, 1976.

Santhi, R. and K M Sellamuthu: *Fundamentals of Forest Soils*, Satish Serial Pub, Delhi, 2008.

Sharma, Rajni: *An Introduction to Plant Morphology*, Campus, Delhi, 2004.

Shubhrata R. Mishra: *Morphology of Plants*, Discovery, Delhi, 2004.

Siddiqi, M.R.: *Tylenchida Parasites of Plants and Insects*, Wallingford, UK, CAB International, 1986.

Singh H. P. and Chadha K. L.: *Genetic Resources of Citrus*, New Delhi, Malhotra Publishing House, 1993.

Singh, Chandra Prakash: *Applied Geomorphology : A Study*, B R Pub, Delhi, 2002.

Sood, M.L.: *Reptilian Nematodes from South Asia*, International, Delhi, 1999.

Sultan Singh and I S Pawar: *Trends in Wheat Breeding*, CBS Pub, Delhi, 2006.

Tawde A. B.: *Propagation and Rootstocks of Mango*, New Delhi, Malhotra, 1993.

Thomas E.: *Fruit Production in St. Kitts and Nevis*, Port of Spain, IICA, 1996.

Tous, J.: *Carob Tree (Ceratonia siliqua L.)*. Rome, International Plant Genetic Resources Institute, 1997.

Urton, Gary: *The Social Life of Numbers*, Austin, University of Texas Press, 1997.

Vanderplank, J.E.: *Plant Diseases: Epidemics and Control*, New York, NY, USA, Academic Press, 1963.

Whealy K.: *The Garden Seed Inventory*, Decorah, Seed Saver Publications, 1988.

Index

L

M

N

O

P

R

S

T

V

W

X

Y

❑❑❑